Aarron Walter

DESIGNING FOR EMOTION

MORE FROM A BOOK APART

Visit abookapart.com for our full list of titles.

Publisher: Jeffrey Zeldman
Designer: Jason Santa Maria
Executive Director: Katel LeDû
Managing Editor: Lisa Maria Martin
Editor: Sally Kerrigan
Copyeditor: Danielle Small
Proofreader: Katel LeDû
Book Producer: Ron Bilodeau

Managing Editor, first edition: Mandy Brown
Editor, first edition: Krista Stevens
Copyeditor, first edition: Rose Weisburd
Compositor, first edition: Rob Weychert

ISBN: 978-1-937557-93-5978-1-937557-94-2

A Book Apart
New York, New York
http://abookapart.com

10 9 8 7 6 5 4 3 2

TABLE OF CONTENTS

For Jamie, Olivier, and Bellamy. Y'all are awesome.

FOREWORD

IT'S NO SECRET THAT our metaphorical web designer toolkit is starting to buckle under its own weight. As more and more skillsets become expected of those who work on the web, none have been more critical in recent years than an understanding of emotional design.

The best thing about designing for emotion is that it doesn't require us to learn new code, frameworks, or hacks. It takes us back to basics and, dare I say, might even make you fall in love with building for the web again.

You're going to learn how gestures of inclusion will turn people into superfans of your work. You'll understand how to acknowledge fear in users and overcome it before they've even had to point out their concerns, and identify blind spots on your team so that your work is some of the most progressive on the web. All things that we *should* have been doing for years, but often leave out for the sake of perceived speed. In recent years it's been seen as a nice-to-have rather than a necessity.

In this era of the web, this is the new standard.

Aarron is *the* leading voice on this subject matter. In this book, he takes his extensive experience in this field and formulates it all into neat takeaways that would otherwise take you years to learn from experience. My advice to you is to treat it like a workbook. Thoughtfully pause at the end of each chapter. Write some notes in the margins and connect your own emotional experiences to the lessons within. You'll gain a personal resonance with the lessons and truly see just how powerful this new skillset is.

If you've ever been lucky enough to cross paths with Aarron, you will read this book and hear his voice, with all the warmth and calmness his energy brings. For those of you yet to meet him, it will be like a long coffee break with a brilliantly smart friend. A true testament to many of the human-to-human lessons inside.

Enjoy—you couldn't be in safer hands.

—Sarah Parmenter

1

EMOTIONAL DESIGN

REVOLUTION: SOMETHING LOST AND SOMETHING FOUND

POWERED BY A CHAIN REACTION of ideas and innovations, a revolution of industry swept the Western Hemisphere in the late eighteenth and nineteenth centuries. In a relatively short time, we discovered ways to transform mined materials into manufacturing devices, transportation systems, and agricultural tools that fueled the twentieth century's explosive modernization. Inventions like the cotton gin, machine tools, the steam engine, the telegraph, and the telephone promised a future filled with technological innovation that produced significant industrial advantages.

Though the Industrial Revolution sprang from a utopian vision of innovation, millions were exploited or enslaved in the process. Industrialization put the machine of innovation first and human needs second.

As the machine found its place in our world, the human hand's presence in the production of everyday objects slowly faded. Factories that could produce goods faster and at a lower cost replaced skilled craftspeople like blacksmiths, cobblers, tinsmiths, weavers, and many others.

But some challenged the myopic march toward industrialization. As mass production expanded in the mid-nineteenth century, artists, architects, and designers founded the Arts and Crafts movement to preserve the artisan's role in domestic goods production, and with it the human touch. The founders of the Arts and Crafts movement revered the things they designed, built, and used every day. They recognized that an artisan leaves a bit of themselves in their work, a true gift that can be enjoyed for many years.

In the present day, we can see a few parallels. A quest for higher crop yields and lower production costs has transformed farms into headless corporations pitting profits against human welfare. But local farmers are finding new markets as consumers search for food produced by people for people. While big-box stores proliferate disposable mass-market goods, platforms like Etsy, Kickstarter, Shopify, and Squarespace are empowering artists, craftspeople, and DIY inventors who sell goods they've designed and created.

Enmeshed in the goods we buy from independent artisans is the human touch—a careful consideration of details that shape the user experience. It resonates palpably, offering evidence of the maker and connecting us on a human scale. There is great power here and a lesson for us as we design digital experiences.

There are plenty of opportunities to build fast and cheap with no reverence for craft or the relationship we have with our audience. Frameworks like Bootstrap make it easy to build from boilerplate but the results, like a mass-produced product, are indistinguishable from others (FIG 1.1). We could reduce our industry to a commodities race, like those who manufactured the Industrial Revolution. There is a market for that kind of work.

Or we could follow a different path, one paved by the artists, designers, and architects of the Arts and Crafts movement, who believed that preserving the human touch and showing

FIG 1.1: As Sarah Parmenter pointed out in her An Event Apart talk "Practical Branding," many websites use boilerplate elements, trading creativity for convenience. The result is that so many websites look the same (http://bkaprt.com/dfe2/01-01/).

ourselves in our work help to form an emotional connection with people that is powerful, unique, humane, and essential.

While the human touch doesn't exist physically in digital design, the ethos of it does. Designing for emotion—creating things that transcend function to engage us on an emotional level—is attainable in all mediums whether physical or virtual.

The dawn of the Industrial Age began with optimism for a modern world but gave way to sober recognition that what we'd created had come at a price. I hear the echoes of our past as we navigate new challenges on the web we're building today. What we've built is equally as revolutionary, but the impact is coming into focus.

THE WAY WE WERE

Like many designers, I had an optimistic view of the web back in 2011 when the first edition of this book was published. Modern mobile devices, faster wireless connections, the rise of new social platforms, and a pervasive entrepreneurial spirit fueled my optimism. It felt like a golden age when design and the web could transform lives for the better.

I thought social platforms would give voice to the marginalized, evolving digital products would empower our creativity, and unfettered access to information would lead to a more democratic and equitable world.

Some of that came true, but like our industrialist ancestors, I was naive to the worst-case scenarios that could arise, and I wasn't alone.

Today, I'm a little more cynical. Grim news of how the web is used nefariously for political or financial gain seems unending (http://bkaprt.com/dfe2/01-02/). Social platforms are, in my judgment, doing almost as much harm as good, and so many of the digital products we use are not inclusive because there's still little diversity in the groups designing and building them (http://bkaprt.com/dfe2/01-03/).

It's time to point ourselves in the right direction. Though our intentions for our work may be good, we need to be aware that the outcomes they shape don't always match our vision. We need to evolve our approach.

In this second edition, we'll build upon the key emotional design principles that guided us back in 2011 and introduce new ones that will help you meet the challenges of today.

We'll get a fresh perspective on designing human experiences filled with emotions—good, bad, and all that's between. If in its optimism, the first edition of this book over-indexed on designing delightful experiences, this edition corrects the balance by looking at more complicated emotions we bring to our work.

We'll also look at how design can be used to create belonging. When we prioritize diversity and inclusion, we reach more people and build lasting relationships with our audience.

Finally, I want you to take away more than just updated principles, but also conversation starters about the business value of emotional design so you can speak with engineers, product managers, executives, and other stakeholders convincingly.

So, where do we start? Well, like any good user experience designer, we begin by understanding the needs of the people we're designing for.

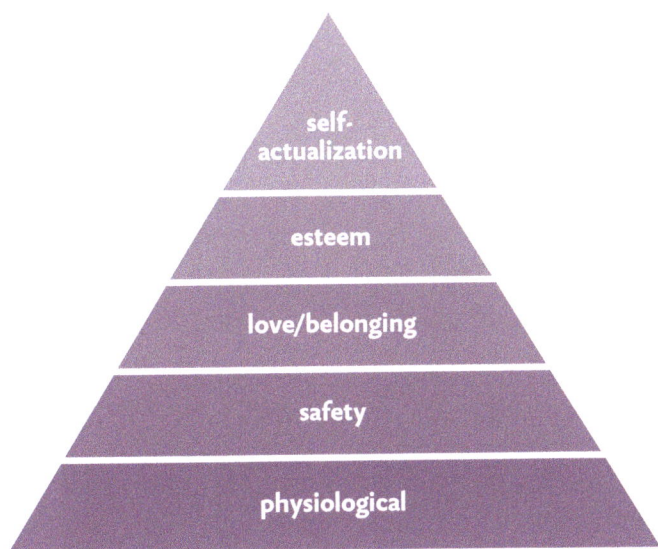

FIG 1.2: Maslow's Hierarchy of Needs.

HELLO, MASLOW

In the 1950s and '60s, the American psychologist Abraham Maslow discovered something that we all knew but had yet to put into words: No matter our age, gender, race, or station in life, we all have basic needs that must be met. Maslow illustrated his ideas in a pyramid he called the Hierarchy of Needs (**FIG 1.2**).

Maslow stressed that the physiological needs at the base of this hierarchy must be met first. The need to breathe, eat, sleep, and answer the call of nature trump all other needs in our life. From there, we need a sense of safety. We can't be happy if we fear bodily harm, loss of family, property, or a job. Next, we need a sense of belonging. We need to feel loved and intimately connected to other humans. This helps us get to the next level: a sense of self, respect for others, and the confidence we need to excel in life. At the top of Maslow's pyramid is a broad, but important category--self-actualization. Once all other needs are

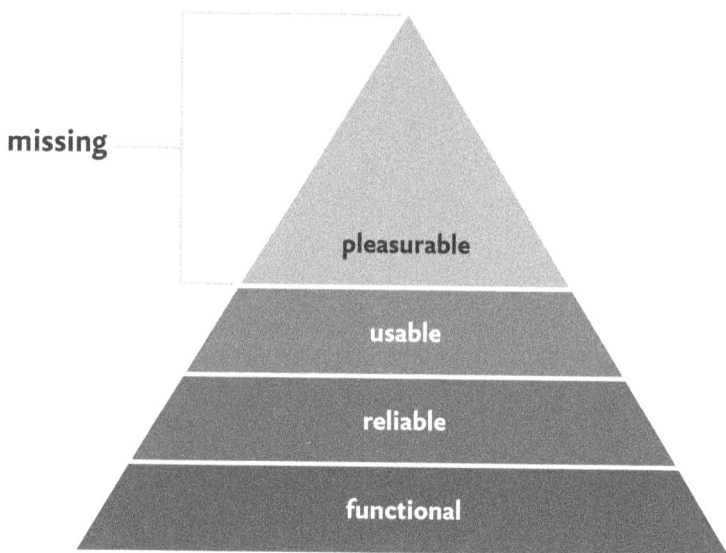

missing

pleasurable

usable

reliable

functional

FIG 1.3: We can remap Maslow's Hierarchy of Needs to the needs of our users.

met, we can fulfill our need to be creative, to solve problems, and to follow a moral code that serves others.

Maslow's approach to identifying human needs can help us understand our goals when designing interfaces. We could be content with only meeting the bottom three strata of the needs pyramid—physiological comfort, safety, and belonging. But it's in that top layer that we can live a truly fulfilled life.

Interface design is design for humans. What if we translated Maslow's model of human needs into the needs of our users? It might look something like this (FIG 1.3).

GETTING THE BASICS RIGHT

Here's how we might remap Maslow's Hierarchy of Needs to the needs of our users:

- *An interface must be functional.* If the user can't complete a task, they certainly won't spend much time with an application. Apple Mail was once the center of my universe, but poor search quality made it impossible to be productive. After years of mastering quick keys, I abandoned it for a mail client that satisfied my basic needs.
- *The interface must be reliable.* Google Hangouts has been eclipsed by competitors because it's just not reliable. Calls drop, image quality is poor; after a couple of bad experiences, it can no longer be trusted for important conversations.
- *An interface must be usable.* It should be relatively easy to learn to perform basic tasks quickly, without a lot of relearning. Ever tried to use Salesforce to perform a basic task? If you have, I'd bet a five-spot that expletives tripped across your tongue a time or two. You're not alone. Though the design team has made significant strides in improving the product, it's hard to make something as complex as Salesforce usable by mere mortals.

Historically, usability has been the zenith of interface design. Isn't that a bit depressing? If you can make a usable interface, you're doing well in our industry. Imagine if we used that yardstick to measure success in the auto industry. By that standard, we'd be swooning over the 1978 AMC Pacer. In response, many websites and applications are redrawing the hierarchy of needs to include a new top tier filled with delight, trust, empathy, and vulnerability. This brings me to the final guideline.

- *An interface must be pleasurable.* Whether it's an offbeat moment that puts a smile on your face or a thoughtful opt-in dialog that inspires your trust, an interface that connects on an emotional level while helping you complete a critical task is powerful. That would be an experience you'd recommend to a friend; that would be an idea worth spreading.

We need a new stick to measure the success of our designs. We can transcend usability to create truly extraordinary experiences.

SURPASSING EXPECTATIONS

Designers creating interfaces that are just *usable* is like a chef creating food that's just *edible*. Think back to the best meal you've ever had. Not a good meal, I mean a mind-blowing, palate-challenging, fall-in-love-with-food-again, great meal. What made it so memorable? Was it the taste and texture of the food? Was it the unexpected pairing of flavors? Was it the artful presentation, attentive wait staff, or ambiance? Chances are, many of these factors worked in concert to elicit an intense emotional response.

Now think about this. Did you once think about the meal's nutritional value? I doubt it. Though the meal met your body's needs, the immense pleasure of the experience formed the memory in your brain, one which you will carry for a long time.

Why don't we aim for a similar target in design? Certainly we want to eat edible foods with nutritional value, but we also crave flavor. And perhaps even ambiance! Why do we settle for usable when we can make interfaces both usable and pleasurable?

Headspace: More than usable

Headspace (headspace.com), a popular app that helps people build mindfulness by developing a regular meditation habit, has emotional design baked into the interface (FIG 1.4). Though the benefits of meditation are well documented, for many of us, it's a challenging thing to do. Headspace makes the abstract concepts of mindfulness accessible through simple animated videos that use metaphors and endearing characters to take the edge off of an activity that can feel daunting.

Press "play" to begin meditation and the circle containing the play button begins to undulate, moving like the mind in unpredictable directions, impossible to control. It's reassuring, as if it's acknowledging that our imperfections are welcome here.

For me, meditation once felt like stepping into a void, but Headspace is full of subtle cues that make the experience feel as if there's a friendly guide pointing the way. The app is functional, reliable, and usable—and so much more. It puts forth a

FIG 1.4: Headspace makes mindfulness accessible with animated videos featuring endearing characters who take the edge off of an activity that can feel daunting.

clear personality that, like a gracious host, makes their users feel comfortable and welcomed.

That persona isn't a thin veneer added at the end of a development cycle. The team that makes the app manifests it through the app copy, the instructions delivered by the warm voice of founder Andy Puddicombe, the colorful, imperfect shapes, and simple animations. That reassuring personality is clearly baked into every aspect of the product, exuding a human quality we can see and feel.

As we'll see in Chapter 4, Headspace uses emotional engagement to create lasting impressions with their customers. There's actually some science behind what they're doing, too. As it turns out, emotion and memory are closely linked.

EMOTION AND MEMORY

Emotional experiences make a profound imprint on our long-term memory. We generate emotion and record memories in the limbic system, a collection of glands and structures under the brain's foldy gray matter. In his book *Brain Rules,* molecular biologist John Medina shares the science behind the relationship between emotion and memory:

> *Emotionally charged events persist much longer in our memories and are recalled with greater accuracy than neutral memories. How does this work in our brains? It involves the prefrontal cortex, the uniquely human part of the brain that governs "executive functions" such as problem-solving, maintaining attention, and inhibiting emotional impulses. If the prefrontal cortex is the board chairman, the cingulate gyrus is its personal assistant. The assistant provides the chairman with certain filtering functions and assists in teleconferencing with other parts of the brain—especially the amygdala, which helps create and maintain emotions. The amygdala is chock-full of the neurotransmitter dopamine, and it uses dopamine the way an office assistant uses Post-It notes. When the brain detects an emotionally charged event, the amygdala releases dopamine into the system. Because dopamine greatly aids memory and information processing, you could say the Post-It note reads "Remember this!" Getting the brain to put a chemical Post-It note on a given piece of information means that information is going to be more robustly processed. It is what every teacher, parent, and ad executive wants.*

Add us designers to that list too, Dr. Medina!

There's a very practical reason that emotion and memory are so closely coupled—it keeps us alive. We would be doomed to repeat negative experiences and wouldn't be able to consciously repeat positive experiences if we had no memory of them. Imagine eating a delicious pile of tacos and not having the sense to eat more the following day. That's a life not worth living, my friend.

A similar feedback loop happens in interface design. Positive emotional stimuli can provide a dopamine hit that encourages repeat engagement with your users. Although the affable design and animation in Headspace's interface may seem like window dressing, it's actually a clever brain hack that builds a positive memory, increasing the chance that Headspace users will learn the concepts it's teaching and make a habit of using the application to meditate.

This is a power that we can use for good or ill; we've become all too familiar with apps that encourage compulsive check-ins and gamify use. It's the flip side of the same principle at work. Headspace has identified where user and business needs intersect and serves both with emotional design. Get to know your customers and what they really need; these insights can help us design profound experiences and create space for emotional design to have the greatest effect.

Intuit: Owning a moment

In 1983, the year Scott Cook founded Intuit, he started a research initiative called Follow Me Home. He dispatched employees to shopping centers and strip malls all over America, where they approached people purchasing one of their personal-finance products and asked if they could follow them home to watch them install the software. Yeah, it sounds a tad creepy—but equipped with proper identification and a good attitude, Intuit employees managed to find their way into their customers' homes and developed a deep understanding of their needs through direct observation of their pain points.

Cook believed the empathy resulting from this practice would be the foundation upon which they could innovate, and although Intuit employees no longer stake out malls to follow people home, they do continue to spend time with customers and have adapted a design ethos that Intuit's former CEO Brad Smith summed up as "Design for Delight" (http://bkaprt.com/dfe2/01-04/).

Delight is a rather myopic way of looking at emotional design, as it's not always the best card to play (something we'll explore more with customer journey mapping in chapter 5).

Personal finance is filled with points where a well-intentioned moment of delight would be most unwelcome. Garron Engstrom, formerly a designer on TurboTax, takes a more nuanced perspective:

> Emotional design is not just about delight and positive emotion. In reality, emotional design is about all emotion—good, and especially bad. If the user is feeling uncertain or fearful, don't shy away from that or sweep it under the rug—instead lean into that emotion. Let the user know you understand where they are emotionally and offer a way to put them at ease. (http://bkaprt.com/dfe2/01-05/)

The design team working on TurboTax translated the ethos of this principle into something they could apply to a broader range of problems. They use the term "ownable moments" to refer to points in the user experience where they can identify a change or escalation in emotion, either good or bad.

One such moment happens in the TurboTax filing process when users are asked about life changes that might influence the filing, including the loss of a loved one (FIG 1.5).

This ownable moment takes deep insight to recognize, but requires minimal effort to write the humane, well-placed piece of copy that profoundly affects customers. Here's how one customer responded:

> I finally got around to doing taxes yesterday. After our information was transferred from last year's return, it asked if either of us had passed away. I entered the information that [husband] died on June 15, and a screen came up that said, 'we're sorry for your loss.' I sat there and stared at it, crying, for a few minutes. It was so cathartic! Please pass on to the team how much that one little sentence meant to me. Whoever thought that up must be a very caring person. (http://bkaprt.com/dfe2/01-05/)

This is but one example of Intuit's "ownable moment" philosophy in action. It remains a core tenant of their culture today.

We'll explore methods for mapping customer emotions in Chapter 4, and how to design your own ownable moments in

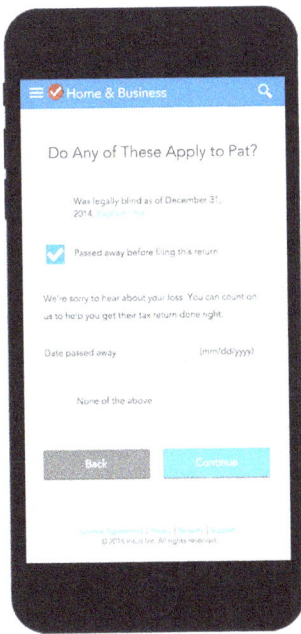

FIG 1.5: TurboTax designers recognized an "ownable moment" in the filing process when users are asked if a loved one has passed. The response is humane and appropriately recognizes the emotional state of the user.

Chapter 5. But first, let's look at the underlying design principle behind Intuit's design philosophies.

THE EMOTIONAL DESIGN PRINCIPLE

People will forgive shortcomings, follow your lead, and sing your praises if you recognize and respond to their emotional state. This is what I call the emotional design principle, and it can be applied in any business.

To engage your audience emotionally, you must let your brand's personality show. When you share your brand's personality, you create opportunities for your audience to relate to the humans who bring it to life. Humans want to connect with real people. We forget that businesses are just collections of people—so why not let that shine through?

Emotional design turns casual users into believers ready to share their experiences. It also offers a safety net that encourages your audience to stay when things go awry. Emotional design isn't just about copy, animations, or design style: it's a different way to think about how you communicate with your users.

Certainly, emotional design has risks. If emotional engagement compromises the functionality, reliability, or usability of an interface, the positive experience you wanted will mutate into a rant-inducing disaster for your users.

We'll touch on the proper use of emotional design. But before we do, let's get comfortable with the firmware that powers the human perspective, as it forms the framework for the strategies you'll craft in your work.

2 DESIGNING FOR HUMANS

WE HUMANS ARE COMPLEX beings and can be difficult to design for. We all have distinct personalities, emotional baggage, and unique dispositions, so how can we design something that appeals to such wide-ranging perspectives?

Beneath disparate personalities and perspectives lie universal psychological principles common to all people. These principles are invaluable tools in our quest to design for emotion. In this chapter, we'll explore the psychological firmware we share and establish a foundation on which we can build emotional design strategies.

THAT WHICH UNITES US

If there is one trait common to all humans, it is that we all emote. In *The Expression of the Emotions in Man and Animal*, Charles Darwin observed:

> *The same state of mind is expressed throughout the world with remarkable uniformity; and this fact is in itself interesting as evidence of the close similarity in bodily structure and mental*

disposition of all the races of mankind. (http://bkaprt.com/ dfe2/02-01/)

What Darwin suggests is that we have a common emotional lexicon guiding us through life. We don't develop emotions by watching others. We're born ready to express pain, joy, surprise, anger, and other emotions. Emotion is an essential survival tool. It's how we communicate our needs to our caregivers, and later in life, it's how we build beneficial relationships. Though we develop verbal language as we mature, emotion is our native tongue from the moment we enter this world. It is the lingua franca of humanity.

HUMAN NATURE AND DESIGN: BABY-FACE BIAS

We can learn a lot about design and how to communicate effectively with our audience by studying evolutionary psychology. As humans have evolved physically, so too have our brains, to naturally select the most advantageous instincts and behaviors that will keep our species alive. We call these instincts "human nature." They're the Rosetta Stone that offers insight into why we behave the way we do. Let's look at a familiar instinct and see how it may inform our design work.

If you're not a parent, you might wonder why people would want to subject themselves to sleepless nights, poopy diapers, and constant caregiving while relinquishing the freedoms and delights of adulthood. On paper, it sounds pretty bad. But in reality, it's pure magic for reasons that are hard to explain.

Shortly before I began writing the first edition of this book, I became a parent for the first time. Since then, my second son was born. Holy cow, is it hard work! But those magical moments—watching them read for the first time, a surprise hug tackle, or just a smile—make me forget that I'm running on three hours of sleep and someone didn't get their pee-pee in the potty (again). All I feel is overwhelming love, and it's totally worth it.

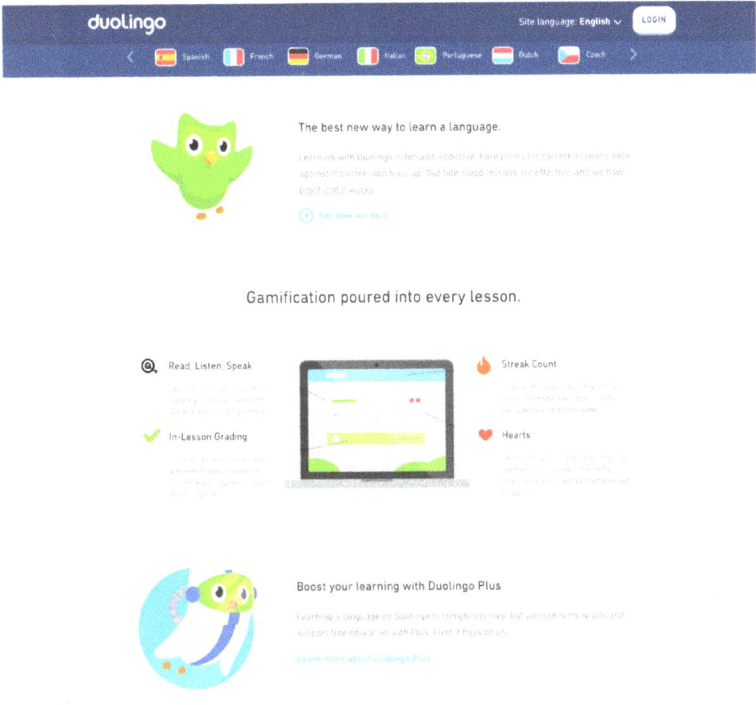

FIG 2.1: Duolingo uses the baby-face bias principle to endear their brand to their customers.

Evolution has given us baby goggles that help us look past the downsides of parenting and trigger waves of positive emotions when we see a little one's face. The proportions of a baby's face—large eyes, small nose, pronounced forehead—are a pattern our brains recognize as very special. Faces that have such proportions are perceived as innocent, trustworthy, cute, and lovable. We're hardwired to love babies.

I know it sounds kind of absurd, but scientists believe we evolved to love babies' faces to prevent us from killing them. Cuteness is a baby's first line of defense. As the late evolutionary biologist Stephen Jay Gould explains in his essay "*A Biological Homage to Mickey Mouse*" (http://bkaprt.com/dfe2/02-02/, PDF), cartoonists have exploited this principle for decades, creating characters with large heads, small bodies, and enlarged eyes that endear them to us.

Designers also use this principle, called the baby-face bias, to their advantage. Can you think of any websites that use a cute mascot to create connections with their audience? There are boatloads of them; GitHub, Duolingo (**FIG 2.1**), and Mailchimp are just a few.

The takeaway here is not to make your website cuter. With a little consideration, you'll discover that behind every design principle is a connection to human nature and our emotional instincts.

THE WORLD IS OUR MIRROR

We project ourselves onto so much of what we see. For example, when we stare at the clouds or inspect the grain of a gnarled piece of wood, we inevitably construct the image of a face in our mind's eye. We are accidental narcissists seeking that which we know best—ourselves.

This instinct is guided by our primordial desire for an emotional connection with others. For this reason, including photos of human faces in your designs can profoundly influence your audience.

We don't have to see two eyes and a mouth to feel an emotional connection to a design. Sometimes we perceive human presence through abstract things, such as proportion. Pythagoras and the ancient Greeks realized this when they discovered the golden ratio (http://bkaprt.com/dfe2/02-03/), a mathematical division of proportions found repeatedly in nature, including the human form (http://bkaprt.com/dfe2/02-04/).

We've used the golden ratio for thousands of years to create art, architecture, and designs that are universally perceived to be beautiful. Though we may not consciously understand that it's present in an architectural structure like the Parthenon or a product like the original iPod, our subconscious immediately sees a pattern of beauty that we know is also present in our bodies. If you've ever read Robert Bringhurst's brilliant book, *The*

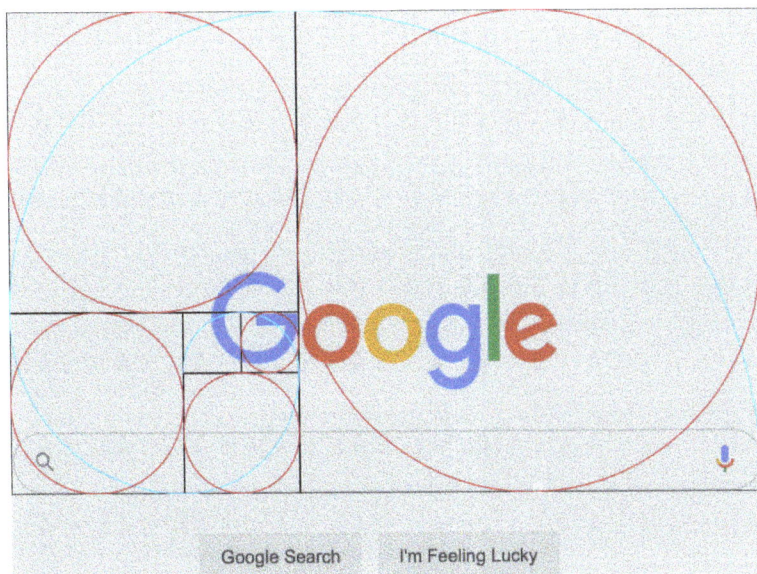

FIG 2.2: The layout proportions of one of the most visited pages on the web, Google's home page, are defined by the golden ratio.

Elements of Typographic Style, you'll know that print designers have used the golden ratio for centuries as the foundation for page layout.

Web designers have picked up on this too. You'll find the golden ratio in the layout of Google's home page and its logo (**FIG 2.2**).

Our ability to find signal and discern patterns in so much noise is a very important trait we use to navigate life, and as you might expect, this ability to recognize patterns greatly affects the way we design.

CONTRAST: IS IT GOOD FOR ME OR BAD FOR ME?

Beyond our ability to express emotion, we also share the instinct to search for patterns. The human mind is beautifully engineered to find differences wherever we look. Our brains constantly scan for patterns in our environment to form insights and keep us from harm.

I experienced this phenomenon recently, as I walked down a pathway on my property at twilight. In a fraction of a second, my brain registered a squiggly line across a circular stone, and I instantly felt a spike of fear. I focused my flashlight on the form to investigate. A young, venomous copperhead stared up at me, poised to strike. I stepped back to steady my beating heart and couldn't help but marvel at how fast my brainstem fired to save me, even before my cerebral cortex could register what I was seeing. It saved me a trip to the emergency room, maybe worse.

We call this break in pattern—strange squiggly line on a circular stone—*contrast*. Our brain's vigilant scan for contrast drives our decision-making process. It helps us decide where we should sit in a classroom (is one seat more advantageous than another?), where we should eat dinner (will one restaurant have better food than another?), and even who we should marry (is one mate more attractive than another?). We use contrast to answer one fundamental question: Is it good for me or bad for me?

We perceive contrast in a couple of ways:

- **Visual contrast:** Difference in shape, color, form, etc.
- **Cognitive contrast:** Difference in experiences or memories

Zebras' patterned hides act as visual contrast to help them blend in with a herd so that stalking lions can't pick them off. When all individuals are high contrast, none stand out.

Humans use visual contrast similarly. Habitual speeders (like me) stick close to other cars to avoid being noticed by police. Speeding alone makes you an easy mark for lurking cops. (I

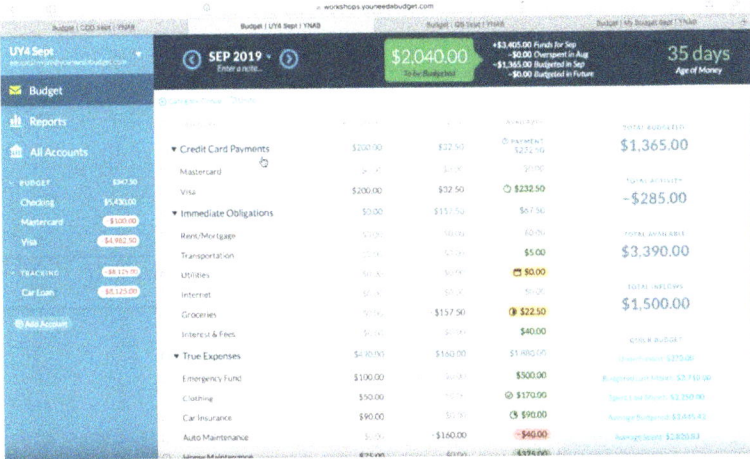

FIG 2.3: Contrast plays a critical role in YNAB's UI to draw attention to overspending.

can't say I advocate this approach, though, as two speeding tickets sit on my desk as I write.)

Police issue speeding tickets to create cognitive contrast in our minds. They hope that the penalties we incur will deter us from repeating mistakes. I can humbly attest that their technique works and has lightened my lead foot.

We can use the power of contrast to great effect in design to make important things stand out. You Need a Budget (YNAB for short)(youneedabudget.com), a personal finance app, uses contrast to direct people's attention to problem areas in their budget.

To manage a budget effectively, you have to know how your money is allocated and when you've overspent. YNAB highlights spending categories that are fully funded with a high contrast green. When you've overspent, they highlight the category in red to let you know you've got a problem to address (**FIG 2.3**). Move money from a category with extra money to cover your overspending, and the red goes away. Financial crisis averted. YNAB's use of contrast makes it easy to understand a complicated budget at a glance.

Remember to consider accessibility as you explore contrast in your designs. The YNAB design team identified hues of red and green that are still discernable to their color-blind users. There are many free online tools that can help you check your work.

Contrast is a powerful tool—but we must take care not to overuse it, as our brains have limitations.

A LIMITED PROCESSOR

Have you ever been to a party where everyone is shouting to speak to the person next to them? As the volume increases, everyone must shout louder to be heard, but that makes it even harder for everyone in the room to have a conversation.

Design works in the same way. As you increase the number of high contrast elements on a page, you proportionally increase the time needed to perform a task, learn a system, and remember pathways. Adding stuff pushes the human brain to its limits. If everything screams for your user's attention, nothing is heard.

Notifications can sometimes make our phones feel like a noisy party. They use the principle of contrast to break our focus with a buzz, on-screen message, or red badges. It's welcome when something urgent requires our attention, but often software designers use notifications irresponsibly, or worse, to manipulate us.

Ecommerce apps that interrupt our lives to tell us, "It's time to shop!" or weather apps that notify us of, "Breaking news! Oldest snow patch in the UK survives!" are using the power of contrast to manipulate people to drive sales and traffic.

It's easy to miss the messages that are actually important in all that noise. Sure, we can turn down the noise in notification settings, but I question the integrity of those who exploit design principles to take advantage of users.

As Mike Monteiro pointed out in his open-sourced essay, "A Designer's Code of Ethics":

By choosing to be a designer you are choosing to impact the people who come in contact with your work, you can either help or hurt them with your actions. The effect of what you put into the fabric of society should always be a key consideration in your work." Mike's code is worth a careful read (http://bkaprt. com/dfe2/02-05/).

Designers will always be pushed by clients and managers to drive business results. But when those requests cross an ethical line or harm users, it's our duty to dissent.

We need to understand how the human mind works to push back effectively. Clients often ask to add more stuff to a page—more navigation items, more products, more promotional content. But the more we add, the more we impair the processing power of our customers' brains.

Hick's Law (http://bkaprt.com/dfe2/02-06/) is a design principle that states that the time it takes to make a decision increases with the number of alternatives available. Though incredibly powerful, the human brain's ability to quickly parse a great deal of information is limited. It's much harder to direct a user to act if their brain has to filter noise. In that respect, we're just like the hungry lioness struggling to find the right zebra to attack.

Every time we add content to an interface or layer on more notifications, it makes it harder for people to identify patterns and contrasting elements. The result is more unpredictable user behavior, poor conversion rates, and lower information retention. (Remind your boss of that the next time you're asked to shoehorn more stuff into your company's homepage.)

But contrast doesn't just shape the way we see things: it also influences our ability to recognize abstract concepts like brands.

BRAND CONTRAST

Branding is all about contrast. When it's done well, it distinguishes a company from competitors, helps drive the growth of the business, and creates a powerful competitive advantage.

As we'll soon see in Chapter 3, brands like Volkswagen, Apple, and Harley-Davidson have used brand contrast to build successful businesses that transcend generations. Their unique brand personas resonate with their customers, inspiring brand loyalty.

Sara Blakely, founder of the successful shapewear brand Spanx, also used brand contrast to differentiate her company amidst a well-established market.

When Blakely started Spanx, the women's undergarment section of department stores was crowded with beige, white, and grey packaging featuring homogeneous models, all with a build that society would deem thin. The competition was both boring and out of touch with the customer. Blakely saw an opportunity for her fledgling business to succeed by doing things differently than the competition.

She designed bright red packaging that stood out in a sea of neutral colors with the tagline, "Don't worry. We've got your butt covered!" It made her brand feel familiar and personal, like she was talking to a close friend at dinner.

Instead of beige packaging with slim models, Spanx packaging featured fun cartoon images of women of various demographics and body types wearing the product.

Most shapewear packaging uses a cardboard insert to uphold the product's form. Where the competition typically made this cardboard disposable, Spanx saw it as an opportunity to make customers smile by adorning it with cartoons inspired by Bazooka bubblegum. The comics humorously showed the flaws in the competition and how Spanx is a far better product. Inspired, some Spanx customers even pinned them up on cubical walls at work or hung them on the fridge.

Traditionally shapewear is meant to be super-discreet, but the tone of Spanx branding made it something to celebrate and a way to challenge body shame culture. Blakley tapped into powerful emotions that helped her build a brand that grew by word-of-mouth.

Spanx used contrast to do something very difficult: They broke into a well-established market and captured the attention and loyalty of customers. Even if you don't use shapewear, you likely know them as a category leader.

Contrast directs our audience's attention to what's important (a warning from YNAB that you've overspent) and what's unique (a brand that's unlike the others). Once we've captured that attention, we can use the powers of design to create a powerful emotional response.

THE POWER OF AESTHETICS

Too often, people see design as the indulgent frosting on a functional interface. Have you ever heard a colleague say, "It would be nice if we could have a slick interface, but people care more about what the site does than how it looks"? Would this person show up to a job interview in their pajamas because people only care about what they can do and not how they look? If they did, they'd discover that thinking is flawed.

Perception is critical. In the coming chapters, I'll show you how design influences emotional engagement and usability.

As Donald Norman, a pioneer in usability and human-computer interaction, points out in his book *Emotional Design,* beautiful design creates a positive emotional response in the brain, which actually improves our cognitive abilities.

> *Attractive things make people feel good, which in turn makes them think more creatively. How does that make something easier to use? Simple, by making it easier for people to find solutions to the problems they encounter.*

Norman is describing the aesthetic-usability effect (http:// bkaprt.com/dfe2/02-07/). Attractive things actually work better. Many brands employ this principle, but none more so than Apple.

Apple's interface design is famously refined, focused, aesthetically pleasing, and usable. Their clean, elegant design makes their hardware and software easy to use—though some, Norman among them, argue Apple has lost its way recently (http:// bkaprt.com/dfe2/02-08/). Apple bakes the aesthetic-usability effect into everything they make, and it keeps their customers coming back.

Apple fanaticism connects directly to their mastery of emotional design. When the late Steve Jobs concluded one of his famous product demos with what became almost a catchphrase, "We think you're going to love it," he truly believed it. It's no mistake that he used the word "love," as their design ethos demonstrates that Apple clearly understands human psychology and emotion.

In 2002, Apple filed a patent for a "Breathing Status LED Indicator." Anyone who owned a Mac in this era will remember the status light on the front of Apple laptops and desktops that gently pulsed to indicate a sleep state. Apple designers considered the context in which this light would most often be seen—in a dark office, a bedroom, or a living room where the status light is one of the only light sources.

The status indicator's pulse rate was very precise. It mimicked the natural breathing rate of a human at rest: twelve to twenty breaths per minute. Like the Parthenon's golden proportions, people might not have made the connection between the light's pulse rate and their breathing rate, but they could feel its calming effect.

Apple could have simply designed the indicator to stay on during the computer's sleep state, and it would have achieved its goal. Instead, their solution was functional and soothing, encouraging their customers to see themselves in the product they used.

A FOUNDATION FOR EMOTIONAL DESIGN

We've discovered some design principles and common traits of the human mind in this chapter, and they'll resurface in chapters to come.

A quick look at evolutionary psychology showed us that much of how we see our world is predisposed at birth, a function of thousands of years of adapting to our environment and finding the best solutions for survival. The baby-face bias is one such example. Contrast also originates in our need to survive, but today we can use it to shape user behavior and make our brand stand out.

We learned that the human mind has limitations. When contrast is overused, we struggle to process our options, as Hick's

Law dictates. And, we discovered that aesthetics is more than just window dressing—it influences usability, as the aesthetic-usability principle illustrates.

This is who we are. We are born with firmware that guides us, and emotion is at the core of that code. Emotion is a fundamental part of who we are as humans, and it plays a foundational role in effective design.

While the human use of emotion to communicate and our reactions to certain situations are universal, designing for emotion still requires nuance and careful consideration. The personalities that sit atop our cognitive firmware make us much more unpredictable. As we'll discover in the next chapter, personality is the platform for our broader emotional responses and the key to making a design more human.

3

PERSONALITY

OUR LASTING RELATIONSHIPS CENTER around the unique qualities and perspectives we all possess. We call it personality. Through our personalities, we express the entire gamut of human emotion. Personality is the mysterious force that attracts us to certain people and repels us from others. Because personality greatly influences our decision-making process, it can be a powerful tool in design.

PERSONALITY IS THE PLATFORM FOR EMOTION

Interface design lives within a broader category called Human-Computer Interaction, or HCI, sitting among the disciplines of computer science, behavioral science, and design. HCI specialists understand psychology, usability, interaction design, programming concepts, and basic visual design principles. Sound familiar? That's awful similar to what user experience designers wrangle every day.

I'll let you in on a secret. I'm not a fan of the name "Human-Computer Interaction." When I design, I work very hard to make the interface experience feel like there's a human on the other end, not a computer. It might sound like I'm splitting hairs, but names are important. Names shape our perceptions and cue us into the ideas that fit within a category.

Emotional design's primary goal is to facilitate *human-to-human* communication. If we're doing our job well, the computer recedes into the background, and personalities rise to the surface. To achieve this goal, we must consider how we interact with one another in real life.

I'd like you to pause for a moment and recall a person with whom you recently made a real connection. Maybe you met them while taking a walk or at an event, or maybe a friend introduced you, and the ensuing conversation was engaging, interesting, and fun. It was their personality that drew you to them, that guided the discussion and left you feeling excited. You bonded thanks to shared jokes, tone of voice, and the cadence of the conversation, dropping defenses and encouraging trust. Personalities foster friendships and serve as the platform for emotional connections.

Hold on to that memory and revisit it when you start a new design project. That feeling is what we're trying to convey through emotional design.

Let's think of our designs, not as a facade for interaction, but as people with whom our audience can have an inspired conversation. We'll create that feeling of excitement and bond with our audience by designing a personality that our interface will embody.

Once again, history can inform our work today. It turns out that designers have been experimenting with personality to craft a more relatable experience for centuries.

A BRIEF HISTORY OF PERSONALITY IN DESIGN

We have a history of injecting personality into the things we make to make mechanical things more human. When Johannes Gutenberg—goldsmith and father of the printing press—exper-

FIG 3.1: Gutenberg's movable type mimicked the calligraphic hand of a scribe in an attempt to make his mechanically-produced bibles feel more human (http://bkaprt.com/dfe2/03-01/).

imented with movable type in the mid-fifteenth century, he was inspired by the human hand. Before the printing press, scribes—usually monks—painstakingly penned each page of religious manuscripts by hand with quill and ink. Transcribing a bible was seen as a sacred duty, as the scribe was thought to be channeling a divine message. For this reason, the hand's presence in these manuscripts had great spiritual importance.

So when Gutenberg designed and cast the original typefaces he used to print hundreds of bibles, the letterforms mimicked the calligraphic style of scribes. Though he created machines to deliver the divine message, he worked hard to make the presentation human (**FIG 3.1**).

We can see the trend of distinctly human design in the twentieth century, when mass production permeated nearly every industry.

The Volkswagen Beetle, released in 1938 and produced until 2003, is the best-selling design in automotive history. Its distinctly human design contributed to its success (**FIG 3.2**). Con-

FIG 3.2: Personality is front and center in the Volkswagen Beetle's design, which helped make it a smashing success through generations (http://bkaprt.com/dfe2/03-02/).

ceived as the "People's Car," the anthropomorphized design makes it more than a car for the people: It's a car that *is* a person. The round headlights resemble eyes above a scoop-shaped hood smiling at us, exemplifying the baby-face bias. Though originally designed for aerodynamics and not personality, the Beetle's "face" conveys a perpetually hopeful and fun attitude that made it easy for generations to connect with, despite dramatic cultural changes over seven decades.

It's hard not to return a smile even if it's coming from an object. That simple design detail constructed an emotional persona for this car, leading to games ("Slug bug red!") and transforming the Beetle into movie hero (Herbie in *The Love Bug*). We've created memories around these experiences, and they remind us of the positive emotions the Beetle inspires.

There is no more concrete an example of personality in design than Apple's "Get a Mac" ad campaign (http://bkaprt. com/dfe2/03-03/, video). Remember that? In the ads, Justin

FIG 3.3: The original Macintosh computer used personality to ease the fears of first-time computer users and to stand out from competitors.

Long portrayed a young hipster Mac who effortlessly tackles complex problems while his rival—John Hodgman's dweeby, uncool PC—bungles every task. These ads conveyed a personality experience and helped consumers compare the differing relationships they could have with their computer. They didn't talk about specs and features; they showed how you would feel if you bought a Mac.

Of course, Apple's design history is filled with examples that illustrate the power of personality. The company anthropomorphized the first Macintosh by using the disk drive to form a charming half-smile (**FIG 3.3**), which helped make computer newbies feel at ease.

The early Macintosh competitors were technical and intimidating. When powering up a PC, users were greeted by an inscrutable blinking DOS prompt. By contrast, I'll never forget the excitement and joy I felt the first time I turned on a Macintosh and was greeted by the now-famous smiling Mac icon. It felt like a real living thing, and I knew then that I'd never go back to a PC. Decades later, I'm writing this book on my Mac. No other brand has captured my loyalty for so long.

Apple, one of the most valuable companies in the world, has shown us time and again that personality can be used as a competitive advantage. It can make products more desirable, memorable, and distinguishable from competitors.

Now that we understand the history of personality in design, let's break these ideas down further to identify how it can create tangible business benefits.

FOUR BENEFITS OF PERSONALITY IN DESIGN

Volkswagen and Apple's thoughtful brand work in those early days is still paying dividends today and has turned their products into cultural icons.

I've identified four business benefits of establishing strong brand personality:

1. **Stand out from the crowd:** A product with a clearly articulated personality stands in contrast to competitors, which creates a marketing advantage. Consumers make purchasing decisions based not just on features, but also on how a product makes them feel. If you're competing solely on functionality it's easier for competitors to one-up you and steal sales.

2. **Create long-term memory:** As we learned in Chapter 1, emotional experiences make a profound imprint on our long-term memory. When we use personality to create emotional resonance in our designs, we help our customers remember our brand and our product. This often leads to word-of-mouth marketing from super fans. It's the stuff companies dream of!

3. **Find your people:** In his book Tribes, Seth Godin tells us, "One of the most powerful of our survival mechanisms is to be part of a tribe, to contribute to (and take from) a group of like-minded people." Companies have used this mechanism to create loyal customers. Harley-Davidson, for example, has a devout following of those who live loud and don't follow the rules. It's not for everyone. They lose some customers because of their personality, but they gain far more. Estab-

lishing your niche connects you to those who will be loyal customers for many years to come.

4. **Convey passion, invite passion:** Your brand's personality communicates your passions, which can translate into forms of self-expression for customers. Harley-Davidson riders see themselves in the brand and express that personality loudly. Every year Sturgis, South Dakota, is flooded with passionate motorcycle riders who come together to celebrate their unique lifestyle, the center of which for many is Harley-Davidson. Lots of companies make motorcycles, but only one makes Harleys.

Powerful stuff. So how do you create a personality for a product or brand? Let's take a look at the tools that will get you there.

PERSONAS

In modern design, we research, plan, and create with our audience's attitudes and motivations in mind. User experience designers interview their audience, then create personas—a dossier on an archetypal user who represents a larger group. Think of personas as the artifacts of user research. They help a design team remain aware of their target audience and focused on their needs.

The persona example shown in FIG 3.4, created by designer Todd Zaki Warfel, tells the story of Julia, a user who wants to visit New York and study at NYU. Through this document, we learn about her demographic details, her interests, her expertise in various subjects, and what influences her decisions on subjects germane to the project; we start to understand who Julia is. This glimpse of her personality helps us understand her motivations and shapes the design decisions that follow.

If you're new to creating personas, the Nielsen Norman Group has a fantastic guide that will help you get started (http:// bkaprt.com/dfe2/03-04/), and if you've struggled to build personas effectively, their video "Why personas fail" (http://bkaprt. com/dfe2/03-05/)will get you back on track.

THE VISITOR

Julia

JUNIOR; MEDIA & COMMUNICATIONS; NON-NYU STUDENT

Goals: Experience NYC and NYU; Take new/different courses; Internship opportunities; Resume builder

Pain Points: Limited courses offered; Costs; Communication issues between home university and NYU in regards to financial aid and loans; Did not have equal access to NYU resources before program; Lack of information about housing

> I had the best time of my life. I loved Spring in NY. I'm visiting my friends in a few weeks. The program itself...I had these great bonding experiences.

Julia is looking forward to doing a semester away at NYU and feels it was fate the way she received a random brochure about the program in the mail. It didn't take too much to convince her parents as they saw that opportunities were available to her in NYC that she would never have at her home school.

Although the application process was a breeze, the communications about housing and getting her loans transferred were a nightmare. If it wasn't for her father following up she probably wouldn't have been able to have the amazing experience she had. Julia signed up for the Self-Designed study to get a mix of classes — searching for classes was quite cumbersome, but she was able to take some great classes that her home university did not provide. Even more exciting was the internship opportunity she landed — an opportunity she would have not had anywhere else. As for the classes and the professors, she expected a little more but the specialized classes were a bonus.

The experience she had made her decide to transfer to NYU full-time. To catch up on requirements, she took a couple accelerated summer classes.

Knowledge

Life Cycle

Activities and Interest

INTERNSHIPS/WORK

EXPLORING

Influencers

messagefirst *design studio*

FIG 3.4: Personas like this one help guide the design process, keeping the focus on user needs.

As we saw in the hierarchy of needs in Chapter 1, we know all users need our designs to be functional, reliable, and usable. By understanding our audience, we can better address their needs. This information also helps us address the top layer in that hierarchy—pleasure—by clueing us in to the design personality most likely to create an emotional connection.

Personas are a standard tool in the design process, but they only provide a partial picture of the relationship we're building with our audience. We know who they are, but who are we? Shouldn't our design have a persona that serves as the counterpart for our user personas to interact with? Why, yes—yes it should.

HOW TO CREATE A DESIGN PERSONA

If your website were a person, who would it be? Is it serious, all business, yet trustworthy and capable? Is it a wise-cracking buddy that makes even mundane tasks fun?

Following a structure similar to a user persona, you can flesh out your design's personality by creating a design persona. Personality can manifest itself in an interface through visual design, copy, and interactions. A design persona describes how to channel personality in each of these areas and helps the web team construct a unified and consistent result. The goal is to craft a personality portrait every bit as clear as the ones Justin Long and John Hodgman embodied in the "Get a Mac" ads.

Before we take a look at a real design persona I created for Mailchimp when I led the design team there, let's examine the components of the document. Here's what you'll include in your design persona:

- **Brand name.** The name of your company, service, or website.
- **Overview.** A short description of your brand's personality. What makes your brand personality unique?
- **Personality image.** Include an image of a person who embodies the traits you wish to include in your brand. This makes the personality come to life. You can choose a famous person, a person your team is familiar with, or a new, ficti-

tious character you've created. If your brand has a mascot or representative that already embodies the personality, use that instead.

- **Brand traits.** List five to seven traits that best describe your brand along with a trait that you want to avoid. This helps those designing and writing for this design persona to create a consistent personality while avoiding the traits that would take your brand in the wrong direction.
- **Voice.** If your brand could talk, how would it speak? Would it speak with a folksy vernacular or a refined, erudite clip? Describe the specific aspects of your brand's voice and how it might change in various situations. People change their language and tone to fit the situation—so should your brand.
- **Copy examples.** Provide examples of copy that might be used in different situations in your interface. This helps writers understand how your design persona will manifest in an interface.
- **Visual lexicon.** If you are a designer creating this document for yourself and/or a design team, you can create a visual lexicon in your design persona that includes an overview of the colors, typography, and visual style that conveys your brand's personality. This would be a helpful addition to any design system guiding.
- **Engagement methods.** Describe the emotional engagement methods you might use in your interface to support the design persona and create a memorable experience. These are your "ownable moments" we learned about in Chapter 1. We'll learn how to create these in the next chapter.

Just as user experience designers post personas in a place where the design, development, and content strategy teams will see them throughout the project, your design persona should be visible and accessible to remind the team of the type of relationship you want to build with your audience. The design persona should guide anyone that crafts a pixel, a paragraph, or a process for your website or product.

Now we'll take a look at some real-world examples. Back when I led the design team at Mailchimp, we created a design persona to guide our work. It proved helpful as our team

mailchimp

expanded to include more designers, writers, and developers and kept us all creating a consistent experience together.

Following the same structure as we've just seen, here's a slightly abbreviated version of the design persona we used.

Brand name: Mailchimp

Overview: Freddie Von Chimpenheimer IV is the face of Mailchimp and the embodiment of the brand personality (FIG 3.5).

Freddie's playful spirit is visible in his kind smile and wink that welcomes users and makes them feel at home. The cartoon style communicates that Mailchimp offers an informal experience that's accessible to all regardless of their skill level. Yes, he's a cartoon ape, but Freddie can still be cool. He's ready with a high five to celebrate your accomplishments at just the right time, and he never gets in the way of your work.

Freddie is casual and creative, inspiring Mailchimp customers to stay true to themselves and do things their own way.

Brand traits: Fun, but not childish. Funny, but not goofy. Powerful, but not complicated. Hip, but not alienating. Easy, but not simplistic. Trustworthy, but not stodgy. Informal, but not sloppy.

Voice: Mailchimp's voice is familiar, friendly, and—above all—human. The personalities of the people behind the brand shine through honestly. The voice of Mailchimp cracks jokes (ones you can share with your mama), tells stories, and communicates with the folksy tone that you might use with an old friend.

Mailchimp uses contractions like "don't" instead of "do not" because that's how real humans speak to one another. Mailchimp uses sound effects like, "hmmmmm..." to make it sound like you're thinking hard, or "Blech, that's awful!" to communicate empathy. Lowercase form and button text reinforce the brand's informality.

Copy examples: *Success message*: "High fives! Your list has been imported." *Error message*: "Oops, looks like you forgot to enter an email address." *Critical failure*: "One of our servers is temporarily down. Our engineers are already on the case and will have it back online shortly. Thanks for your patience."

Visual lexicon: *Color*: Mailchimp's bright yet slightly desaturated color palette conveys a sense of fun and humor. The colors feel refined—not romper room-y. Mailchimp is fun, but it's also powerful. *Typography*: Mailchimp is easygoing, efficient, and easy to use, and its typography reflects it. Simple, sans-serif headings and body copy vary appropriately in scale, weight, and color to communicate information hierarchy, making Mailchimp feel like a familiar, comfortable cardigan that is both functional and beloved. *General style notes*: Interface elements are flat and simple, keeping things easy to understand and unintimidating. Soft, subtle textures may appear in places to warm up the space and make it feel human. Freddie should be used sparingly. He never gives application feedback, stats, or helps with a task.

Engagement methods: *Surprise and delight*: In peak moments in the user experience, like after sending an email campaign, Mailchimp creates a sense of delight with an animated Freddie high five. Easter eggs create unexpected moments of humor that may convey nostalgia through an 8bit video game or reference kitschy pop culture.

The Mailchimp brand continues to evolve along with the product. The Mailchimp team has taken the original design persona much further by creating a public page describing how the personality manifests in every aspect of the brand's design (**FIG 3.6**).

While we were developing that original design persona, Kate Kiefer Lee and team developed Mailchimp's voice and tone guide (http://bkaprt.com/dfe2/03-06) that greatly expanded our

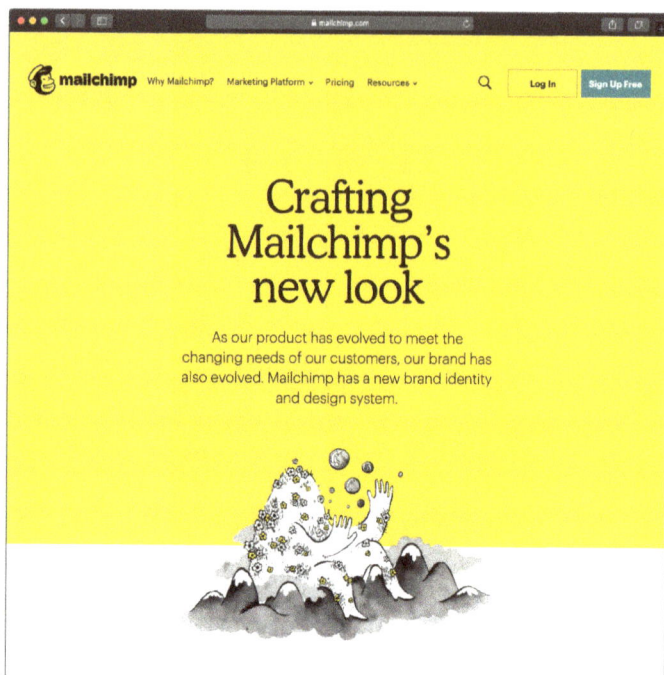

FIG 3.6: The Mailchimp team created an amazing public summary of how the brand's personality translates to their product and marketing design.

understanding of how the brand personality should come through in our writing. Not only did that voice and tone guide help our entire company deliver a consistent brand personality, it also inspired countless teams around the world to build similar guides.

TAKING THINGS FURTHER

Now that you're familiar with the core elements of a design persona and you've seen how it can expand into more detailed guides, I encourage you to consider how you can take it further. A design persona combined with a design system and a voice and tone guide would create an amazing reference point for any team and make onboarding new designers more efficient.

If you want to try this out for yourself, I've shared a design persona template online (http://bkaprt.com/dfe2/03-07) that you can use in your next project.

PERSONALITY IN THE WILD

Many websites are already using personality to shape the user experience and power their commercial success. Though they address different markets and have different business goals, goodr and Headspace have each discovered that personality is the key ingredient in the emotional connections they're building with their audience and in their overwhelming success.

Goodr: Words matter

Goodr sells inexpensive sunglasses. Then again, so do most drug stores. But goodr does it, ahem, goodr.

Their products have names like "Whiskey Shots With Satan" and "Jorts for Your Face," each with a gonzo origin story and product descriptions that entertain as much as they inform (FIG 3.7).

Sunglasses are the product, but what they're selling is an experience—the experience of laughing as you click "add to cart." The experience of sharing a smile with your spouse as you read the brilliant writing in the packaging of your new sunglasses. The experience of having a friend compliment you on your shades at the pool and replying, "these things are called 'jorts for your face.'" Ridiculous, right?

Maybe not.

Significant Objects: Measuring the value of emotion

Writers Joshua Glenn and Rob Walker had a hypothesis that stories are a powerful driver of emotional value that—in theory—should be able to be measured objectively. To test their

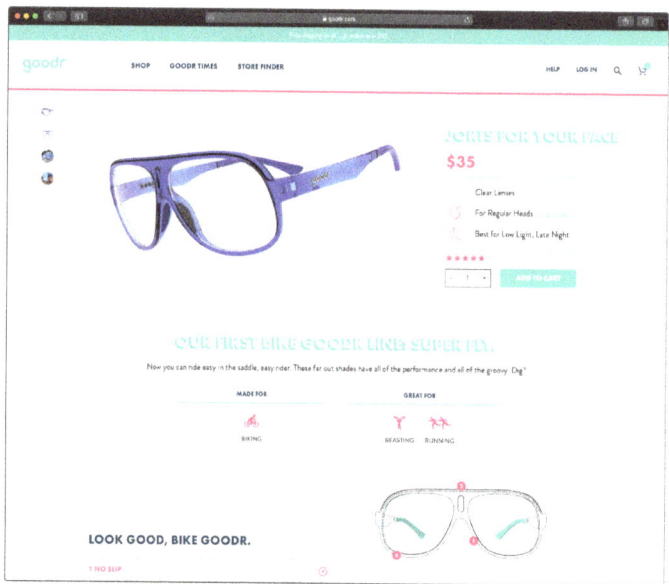

FIG 3.7: Everybody needs jorts for their face, right? Goodr thinks so. Personality and offbeat storytelling drive sales and create devout customers for them.

hypothesis they created the Significant Objects Project (http://bkaprt.com/dfe2/03-08/). They bought $128.74 of what they called "thrift-store junk," wrote compelling stories for each object, then listed them on eBay and sold them for $3,612.51, all of which was donated to charity. They created a 28x return on their investment through storytelling that brought each object to life and created emotional value people were willing to pay for. People buy things because of how they make them feel, as much as, sometimes even more than, because of what they do.

Goodr is doing something very similar: making a pedestrian object more valuable by building a story around it. And those stories are all shaped by strong brand personality.

Headspace: Using personality to reshape perceptions

Back in Chapter 1, we got a glimpse of how Headspace uses emotional design to help people understand meditation. If we look closer, we see that personality is doing most of the heavy lifting. By studying their audience, the Headspace team realized that misperceptions about mindfulness were some of the biggest barriers to getting people to adopt a meditation practice.

For many, meditation conjures serene images of a monk on a mountaintop sitting still as a result of mastering their body and mind. But in reality, meditation is not about perfection. It helps us observe our noisy mind so we are more equipped to deal with life's problems.

Anna Charity, Headspace's former head of design, helps us crack open the personality of the product to see some of the elements behind it and how they're used to teach people about meditation.

> The mind is a complex place, and it isn't always an easy place to inhabit (which is why meditation is so valuable). We knew we had to develop a style that translated these ideas in an approachable and relatable way. Animation and illustration became integral to the brand. By using characters and storytelling, we could break down the barriers of a tough subject matter and present it in a light-hearted but sensitive way. Characters are a great vehicle to represent the weirdness inside your head; it feels playful and memorable as result. (http://bkaprt.com/dfe2/03-09)

Characters and animation convey so much personality. Like our minds, these characters are weird. They're floppy and have goofy expressions and entertaining hairdos—not exactly monkish. Their shapes are curvy and soft, making them almost cozy. The vibrant color palette uses plenty of primary and secondary colors. It could come off as juvenile, but the Headspace design team desaturated the colors to make them feel a little wizened, like a well-loved toy that's seen too much sun time.

FIG 3.8: Headspace worked with talented illustrators and animators, like Bee Grandinetti, to develop characters that not only educate people about the concepts of meditation, they also channel brand personality.

Strangely, these characters still feel sophisticated enough to help us meditate on the heavy issues in our life without making them feel silly, carefully walking the line between disarming our inner skeptic with cartoons while retaining a sophistication that keeps trust intact.

There's a lot going on here. Color, shape, animation, illustration: it all works together to build a personality that makes this company relatable, inviting, and unique.

The right amount of personality

Like a pinch of salt, personality enhances the flavor of a user experience. But use too much and your audience will find it overpowering. Finding the right balance is important.

When we strip the user experience down to the bare essentials, like copy alone, we can see clearly when personality enhances or hinders. Slack provides a helpful reference.

Slack's API lets developers build chatbots, programs that users can interact with through text alone. Recognizing that personality is an important part of the Slack brand and chatbot

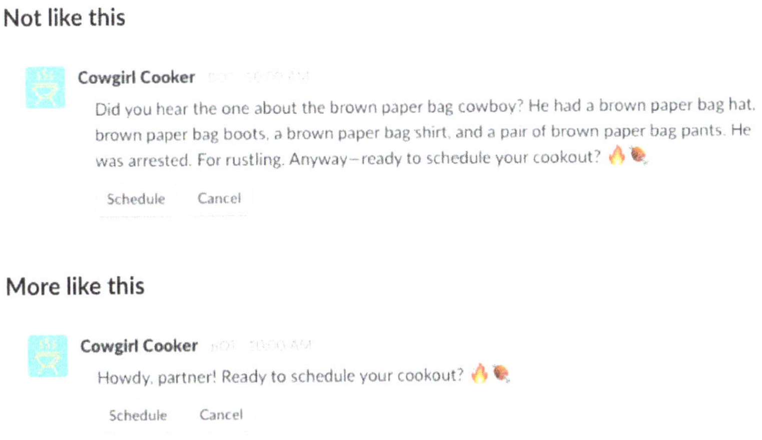

Not like this

Cowgirl Cooker

Did you hear the one about the brown paper bag cowboy? He had a brown paper bag hat, brown paper bag boots, a brown paper bag shirt, and a pair of brown paper bag pants. He was arrested. For rustling. Anyway—ready to schedule your cookout? 🔥 🍖

Schedule   Cancel

More like this

Cowgirl Cooker

Howdy, partner! Ready to schedule your cookout? 🔥 🍖

Schedule   Cancel

FIG 3.9: This example from Slack illustrates the right and wrong way to construct a chat bot and shows us how to convey personality while preserving clarity.

design in general, Slack created a guide to help developers use it without going overboard (http://bkaprt.com/dfe2/03-10).

A bot with personality can bring warmth to an otherwise cold interaction with a computer program, but personality shouldn't get in the way of helping the user complete tasks. As Slack points out in their guide, if your bot's persona is causing you to sacrifice clarity for cleverness, you're laying it on too thick.

> *Don't construct a personality that means you have to add sentence upon sentence in order to get to a joke 'in keeping' with your bot's sense of humor. No one cares. Get to the point. (http://bkaprt.com/dfe2/03-10)*

When personality gets in the way of functionality, we're doing it wrong (FIG 3.9). As we learned in Chapter 1, our users need functional, reliable, and usable experiences. The pleasure that a persona brings to the user experience should never compromise those foundational ideas.

Personality is a big part of what makes Slack feel fun and relatable. As their developer guide illustrates, a little goes a long way.

THE POWER OF PERSONALITY

Personality is a uniquely powerful design tool that can make brands memorable and the experience with their products exceptional. A design persona will help you define the personality traits of your brand or product and will serve as a reminder of when those traits go too far. As Slack's API guide showed us, it's important to know where the edges are.

Just as our personalities shift with the context of communication in real life, they should shift in the projects we design. There's no one-size-fits-all solution. If we stop thinking of the interfaces we design as mere control panels and think of them more as people our target audience wants to interact with, we can craft emotionally engaging experiences that make a lasting impression.

Like personality, empathy and inclusion help us build connections with the people our work serves, but on a level deeper than we've yet seen. In the next chapter we'll see how designing inclusively is essential to the practice of emotional design.

4
EMPATHY AND INCLUSION

EMPATHY—THE ABILITY TO understand and share the feelings of another—is a practice all designers should cultivate. We designers often feel confident in our understanding of users because we invest time researching their behaviors and motivations. But despite our well-intentioned focus on users, there are times when we aren't as empathetic as we'd hoped because we haven't thought inclusively about who might use the things we're designing, and under what circumstances.

Inclusive design is inherently tied to emotional design. When our design decisions leave large groups of people out of the loop, we create feelings of alienation and disempowerment. But when we design inclusively for all, we can create deeply positive experiences that resonate with people—all people—as we'll see later in this chapter.

Inclusive design requires more than our good intentions. We need a commitment to the practice of empathy, and a framework to guide our work.

FRAMEWORKS FOR DESIGNING INCLUSIVELY

Boyuan Gao and Jahan Mantin, co-founders of Project Ink-blot, have created a framework that can help you build diverse perspectives into the creation process (http://bkaprt.com/dfe2/04-01/).

Their framework, called Design for Diversity or D4D for short, has a simple series of core questions that can easily be folded into your design process. Let's walk through each one to see how they shift our perceptions.

Question 1: What's the worst-case scenario, and on whom?

Designing for inclusion requires us to slow down a bit to think through the implications of our design decisions. Gather your design team, key partners, and stakeholders to have a discussion about the worst-case scenarios your work could lead to and who will bear the brunt of the impact.

You'll need to understand these two concepts to make your discussion productive:

1. **Intention:** What you and your team hope to achieve for your customers.
2. **Impact:** How actual communities experience your product or interface.

Many of us only consider our intention and naively believe that the impact of our work will match our vision. Unfortunately, that's not how things work. Let's look at a real example that illustrates how easy it is for intent and impact to misalign, creating serious consequences.

Airbnb recognizes that, for their business model to work, they need hosts and guests to trust each other. One way they foster that trust is by letting each party see a detailed profile with a name, past reviews, ID verification, and a profile photo. Hosts see these profiles when a booking request is made and can choose to accept or reject.

Profile photos are a nice way to humanize that first interaction. Seeing a person's face can help build empathy, or at least that was the intent. The impact, however, was very different.

Racial biases, conscious or not, slip into the guest vetting process when photos and names are included in profiles, and it results in Black guests being denied bookings at an alarmingly high rate. A study conducted by Benjamin Edelman, Michael Luca, and Dan Svirsky found that on Airbnb, "applications from guests with distinctively African American names are 16% less likely to be accepted relative to identical guests with distinctively white names." (http://bkaprt.com/dfe2/04-02, PDF).

To their credit, Airbnb has made changes to their guest vetting process to reduce bias by hiding guest profile photos until after they've been accepted by the host. Airbnb leadership has also built teams who focus on addressing racial discrimination on their platform. We'll learn more about how Airbnb is trying to address trust between hosts and guests in Chapter 6.

There are many good people with good intentions at Airbnb, yet this example reminds us of the complex problems we're designing for today and how essential it is to pause and consider how our designs could be used to hurt people.

Always consider the worst-case scenarios in your work.

Question 2: How do the identities within your team influence and impact your design decisions?

The second question in Gao and Mantin's framework helps us consider how the identities of the people on your team may inform or influence your perspectives as you design. As you might expect, teams with people who have diverse backgrounds and perspectives are advantageous when designing experiences that serve diverse audiences—something to keep in mind in your hiring process.

If your teammates feel safe sharing personal information, it's helpful to have open conversations about the identities each team member brings to their work, to help you see the breadth of perspective and the gaps you may need to address. Gao and Mantin recommend we start that conversation with questions like these:

- What race(s) and gender(s) do you identify as? What other ways do you identify that are important to you? (e.g. queer, Latinx, middle-class woman in her 30s, aunt, grew up in a military family, etc.)
- How might these identities influence and/or inform how you design products or services?
- What perspectives or lived experiences might be missing from your team?

Gaps in your team's perspective will be clear after conversations like this. Once you've identified your team's blind spots, consider ways to bring those perspectives into the light; perhaps by conducting user research with people whose perspectives aren't represented on the team, or better yet, by hiring from more diverse candidate pools. Consult engineering leader Tessa Ann Taylor's guide to building a diverse team to get started (http://bkaprt.com/dfe2/04-03/).

Question 3: Who might you be excluding?

One simple question can help you and your team consider inclusion each step of the way: "Who might we be excluding?" Asking this question in design reviews, meetings, and in those quiet moments of design flow will help you challenge assumptions and consider the impact of your design decisions on all people.

In their framework, Gao and Mantin help us look deeper at exclusion with a brainstorming tool they call "All people" statements. When we frame our assumptions by saying "All people can express themselves in their messages using emojis," we can begin to expose flaws in our thinking and help us see who tends to be excluded from experiences like the one you're designing. Is your "all people" statement true? Are there groups who don't fit into that statement?

You and your team can start to poke holes in your statement to see the communities of people for whom your statement is false. When you see those flaws, you'll understand where you need to go further to include those who have been left out of your design process.

Question 4: How will you engage the people you want to reach within your design process, equitably?

Gao and Mantin point out that when we engage communities who have been exploited and have historically had less decision-making power, it's important to invite them into your process in a way that acknowledges you are thinking about how they can benefit.

That benefit can take many forms: financial compensation, a byline, a role on an advisory board, decision-making power, etc. Your goal should be to create an equitable partnership.

Start by asking what a win for them would be and carefully listen to the response. Here's the example Gao and Mantin recommend:

> *If you're open to sharing, I'd like to learn more about you and your interests and invite you to be a leader on this project. Would you be interested in this, and conversely, do you have any concerns? What would make participating in this project a win for you?* (http://bkaprt.com/dfe2/04-01/)

Your team must work with broad segments of your audience to ensure you're designing inclusively, keeping in mind not just your target audience but also a group Gao and Martin refer to as the "Impact Source." They are people who you may not intentionally be targeting, but who might interact with or be affected by the things you're creating.

The questions from this framework will help to bring these groups more clearly to the surface, and prompt honest conversations that can help us more carefully consider the gaps between our intentions and the impact of our work.

And theirs is not the only framework for inclusive design. Microsoft has created their own toolkit (FIG 4.1) of inclusive design thinking activities, along with an inclusive design primer filled with useful guidance (http://bkaprt.com/dfe2/04-04/).

Maybe you already use some of these tactics, and that's great. The wonderful thing about frameworks is they're meant as starting points, not as dictates. Your designs will end up reso-

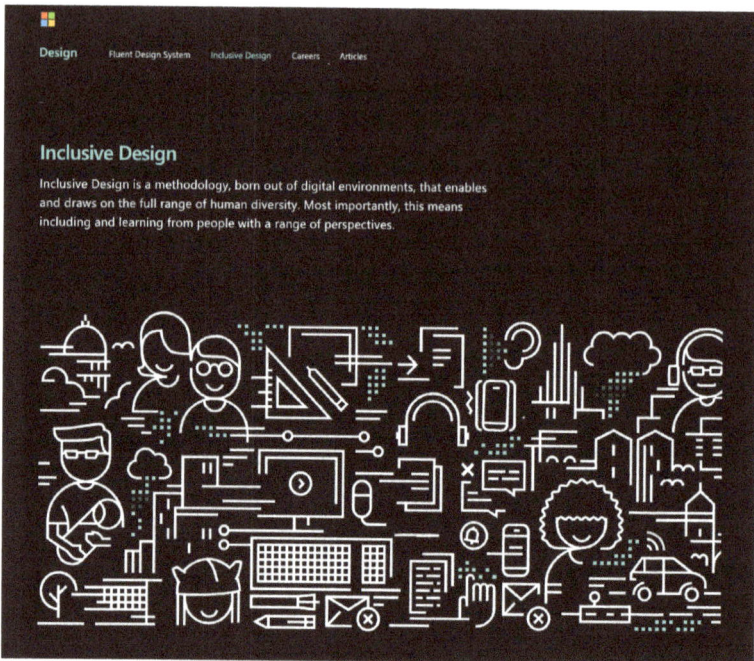

FIG 4.1: Microsoft's inclusive design tools provide insightful activities and principles to help broaden perspectives.

nating with more people once they're able to respond to the questions the frameworks pose.

Now, let's add another tool to your toolbox to help cultivate empathy for your users and allow you to see the world from their perspective.

EMPATHY MAPS

In their book, Gamestorming, authors Dave Gray, Sunni Brown, and James Macanufo introduced a simple and effective way to put yourself in your customer's shoes with a tool called empathy maps (**FIG 4.2**).

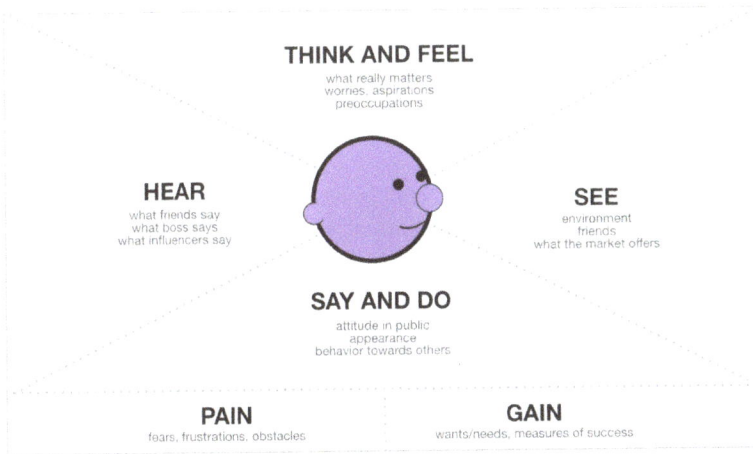

THINK AND FEEL
what really matters
worries, aspirations
preoccupations

HEAR
what friends say
what boss says
what influencers say

SEE
environment
friends
what the market offers

SAY AND DO
attitude in public
appearance
behavior towards others

PAIN
fears, frustrations, obstacles

GAIN
wants/needs, measures of success

FIG 4.2: Creating an empathy map can help your team think and feel from your customers' perspective and help you understand how to design for more complicated emotional experiences.

Creating an empathy map is not a rigorous, research-intensive process like user personas, which we touched on in Chapter 3, but it will help you and your team begin to think carefully about the people who will use the things you're creating.

To create an empathy map, spend some time talking to your customers to learn what's swirling in their minds. This doesn't need to turn into a major project. A few phone calls with customers will give you plenty to work with.

With a bit of user research on hand, you can gather your team to create an empathy map. Draw a face in the center of a large sheet of paper. Label areas around the head: "think and feel," "see," "hear," and "say and do." Draw a horizontal line at the bottom and create "pain" and "gain" labels.

Ask your team to describe the following from the point of view of your customer:

- **Hearing:** What are they hearing from friends, colleagues, and influencers?

- **Seeing:** What are they seeing in the market or their environment? What are they seeing their friends do?
- **Saying and Doing:** What is this customer saying and doing? What behaviors are they exhibiting that may help us understand their motivations?
- **Thinking and Feeling:** What really matters to this person? What are their worries and aspirations?

Finally, at the bottom, work with your team to identify:

- **Pains:** What current pain points does your customer currently feel—fears, frustrations, and obstacles?
- **Gains:** What does this customer want to achieve or gain? What are their measures of success?

Empathy maps push us to consider a user's environment, inputs they're receiving from their community, and emotions (positive and negative) they may experience when using the things we design. In concert with an inclusivity framework like the ones we explored earlier, it's a great tool for cultivating a regular practice of empathy in design.

Now that we know how to modify our process to design more inclusively, let's see how inclusivity has been applied to help underrepresented groups feel a sense of belonging.

BELONGING

As the father of two Black boys I'm acutely aware that pop culture imagery is rarely inclusive, especially in the United States. The protagonists in movies, books, advertising, magazines, websites, and software are white more often than not. If you're a person of color, a lifetime of rarely seeing your reflection in pop culture creates a damaging narrative: you're on the outside.

There are myriad devices we can use to create emotional engagement, but none is as fundamental and powerful as inclusive thinking. Remembering to ask, "Who are we leaving out? Who is not represented in our photography, icons, illustrations, and videos?" can help you snap out of your narrow perspective

Kaya Thomas ✓
@kthomas901

really appreciate the brown hand in this graphic, I'm so used to seeing 'flesh' colored hands in graphics @SlackHQ

Add to Slack

4:53 PM · Aug 25, 2015 · Twitter for iPhone

FIG 4.3: The brown hand in the "add to Slack" marketing campaign created a strong sense of belonging with people of color who have otherwise grown accustomed to seeing White as default.

of your life experience to see the experiences of others. When we are inclusive in our representations of people, we promote a sense of belonging that resonates deeply.

Slack thinks about these sorts of things. When they launched a new way for developers to connect to Slack with a simple "add to Slack" button, Diógenes Brito was the designer tasked to create the marketing images. Under a time crunch, Brito grabbed an existing hand illustration created by a colleague and combined it with the new "add to Slack" button.

He made one key tweak: He changed the skin color of the hand to a shade of brown similar to his skin (FIG 4.3).

Though it was a small design change, a debate raged in Brito's mind as he considered it. He wondered if it was worth the hassle, especially since he had to ask a white colleague who created the source file to make the change for him. He didn't want his request to seem like "a thing." But it kind of *was* a thing, the *right* thing, and he as the designer had the power to make the decision.

So they changed it, they launched, and it resonated; people of color felt a rare moment of inclusion.

Afterwards Brito wrote:

> *Why was the choice an important one, and why did it matter to the people of color who saw it? The simple answer is that they rarely see something like that. These people saw the image and immediately noticed how unusual it was. They were appreciative of being represented in a world where American media has the bad habit of portraying White people as the default, and everyone else as deviations from the norm. (http://bkaprt. com/dfe2/04-05/)*

Decisions like Brito's have even more impact when they're made as part of an identity system that's used as the foundation for all design work in a company. Jennifer Hom, experience design manager of product illustration at Airbnb, had the unique opportunity to rethink the illustration system used in their products and marketing.

She recognized it was an opportunity to make their design practice more inclusive while also aligning to Airbnb's company mission (http://bkaprt.com/dfe2/04-06/):

> *Our mission is to connect people across cultures and continents, so our illustrations have to reflect the community we're bringing together. We can't generalize or curate based on what's considered most 'commercially acceptable.' Instead, we need to be radically honest. Humans are humans, so we don't try to design our way around the simple fact that we're all different. Western-centric, outlined cartoon people need to change. Diversity of age, race, disability, religion, orientation, and gender are the foundation of who we are. We need to reflect belonging anywhere, and my role was a clear opportunity to celebrate diverse identities.*

Hom considers varying body types, genders, disabilities, ages, and multi-racial families, and she depicts interactions

FIG 4.4: Jennifer Hom led the redesign of Airbnb's illustrations system and made diversity the new normal in their design work.

between all peoples. She's made it easy for any designer at Airbnb to be inclusive simply by working with the building blocks and principles she's outlined.

Earlier in this chapter, we learned how, despite their good intentions, Airbnb's vetting process resulted in discrimination against their Black customers. Hom's work on the Airbnb illustration system is just one example of the investment the company is making to design inclusive experiences. They don't always get it right, but we know that pushing biases out of the shadows into clear view puts us on the right path of thinking and designing more inclusively.

By using design to make our customers feel seen, we create deep emotional engagement. When people feel a sense of belonging, they'll give you repeat business and they'll tell others. Designing for inclusion is the right thing to do, and it's also good for business.

With a new understanding of how to cultivate empathy for our users and how we can design more inclusively, let's now consider the timing of emotional design and some principles that will help us design memorable experiences.

5 EMOTIONAL ENGAGEMENT

AS WE'VE SEEN IN Chapter 3, personality is a powerful way to engage your audience. It helps people understand who you are and shapes how they interact with you, while setting the tone for the voice, aesthetic, and interaction design of your site or product. It's a foundational layer on which you can build. And in Chapter 4, we learned just how crucial it is to understand and empathize with the people you're designing for.

Now we're ready to layer moments of emotional engagement on top of that foundation, using the power of psychology to create positive, long-lasting memories of your brand in the minds of your audience.

Before we dive into emotional engagement, though, let's first talk about timing, as knowing when to use emotional engagement is every bit as important as knowing how.

DESIGNING MOMENTS

Perhaps you've heard the term "customer journey?" It refers to the experience a user has with your site or product to reach an intended outcome, like completing a purchase or setting up an

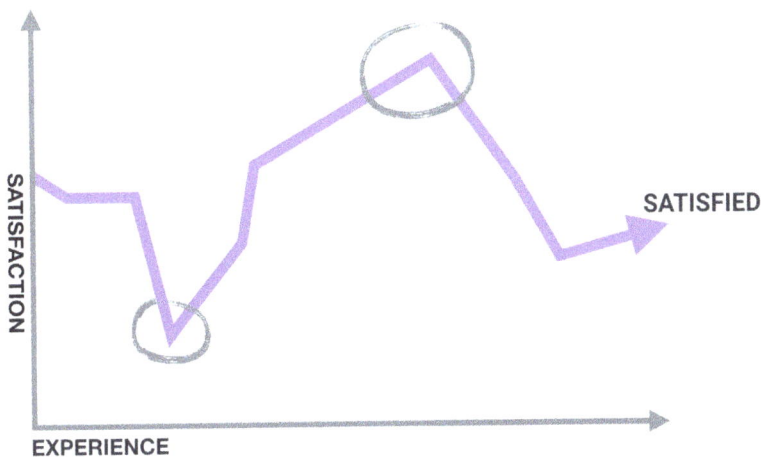

FIG 5.1: By mapping the customer journey, we can see points at which the experience is at its best and worst. When we use emotional engagement at just the right time, we can create a better experience.

account. A customer journey can be more than just a session on your website. It can encompass many sessions online, across devices, or, if you have a brick and mortar store, in person too.

We'd like each step of the journey to be perfect for our customers, but if we're honest with ourselves, there are peaks and valleys. You can see these for yourself, in fact, with a tool called a customer journey map (FIG 5.1). This map will help you identify when to use emotional engagement.

Creating a customer journey map is a team-based activity that goes beyond what we can cover in detail here (though your personas from Chapter 3 will likely come in handy). I recommend this step-by-step guide from the folks at Atlassian to help you get organized (http://bkaprt.com/dfe2/05-01).

The map's peaks and valleys show the various high and low points of the customer experience, with the dotted horizontal line indicating baseline satisfaction and emotional neutrality. When you find valleys, you can take away the pain by improving usability and shortening time to task completion.

Generally, this is the worst time to introduce a witticism from your cheerful mascot, as people are operating within the lowest levels of the hierarchy of needs we saw in Chapter 1. They need functionality, reliability, and usability first and foremost. Do some research to find out what's causing user frustration and make the necessary changes to mitigate it.

Now to those peaks. This is where great things happen, and when emotional engagement is most powerful.

To help you understand the opportunity, let's talk about colonoscopies. Bet you didn't see that one coming! Of course, most people don't. (Rim shot!)

The Peak-end Rule

Back in the mid-1990s, Daniel Kahneman and Amos Tversky conducted a study with two groups of patients who had to undergo an uncomfortable colonoscopy procedure. Patients rated their pain during the procedure, which resulted in a graph not unlike the customer journey map we saw earlier.

For group A, the procedure lasted the standard amount of time, but for group B the procedure lasted an extra three minutes. During these last minutes, the scope remained unmoved, which caused discomfort, but wasn't painful.

Afterwards, the doctors asked all patients how bad the procedure had felt to them. For group A, their memory of the procedure was far worse than for group B, even though their procedure was shorter.

Dr. Kahneman explains that our brains process events through what he calls "the experiencing self" and "the remembering self." The experiencing self processes events in the moment, but not all of those events get committed to memory. Remember in Chapter 1, we talked about the limbic system that records emotional experiences in memory? It's at work here. The remembering self is a storyteller that selectively recalls those memories.

As Dr Kahneman describes in his TED talk on his research (http://bkaprt.com/dfe2/05-02/, video), "endings are very, very important." The remembering self relies most heavily on the end moments. The last thing group A recalls is extreme pain. The last thing group B recalls is mild discomfort.

This is the peak-end rule. The peak at the end of an experience—whether it's positive or negative—will disproportionately shape our memory of the whole experience.

Keeping this principle in mind, when possible, we should position peak experiences towards the end of the user journey to leave a lasting positive impression. It's okay to have peak experiences elsewhere in the customer journey, of course, but mind the peak-end rule to achieve maximum impact.

Let's take a look at a peak-end experience that has had a lasting impression on customers for many years.

A well-timed surprise: The Mailchimp high five

Back in 2005, when I was still a freelancer using Mailchimp to help my clients, I remember spending hours writing and designing emails in the product. When my work was complete, and I pressed "send" releasing thousands of emails to customers, I literally said out loud to myself, "It's Miller time!"

It was a triumph, and I wanted to high-five someone.

So when I joined the company in 2008 and founders Ben Chestnut and Dan Kurzius gave me and my colleague Chad Morris the freedom to redesign the product, I wrote a bit of copy for the page the user sees after they send an email campaign: "High fives! Your email has been sent."

Fast-forward a few years. The team had grown and we were redesigning the product once again. We took a close look at that page, recognizing this was where our customers felt most emotionally engaged. We wanted to deliver on our promise of a celebratory high five!

We used the principle of surprise to shape the moment. In Chapter 2, we learned that our brains scan for pattern breaks to identify contrasting visual and cognitive elements so that we can react appropriately. When we're surprised, we're experiencing a high-contrast situation in which something is not as we expected. A moment of surprise frames our attention, blurs peripheral elements, and brings the extraordinary into focus.

High Fives!

Your campaign is in the
send queue and will go out shortly.

FIG 5.2: The Mailchimp high five that customers experience after sending an
email uses the peak-end rule to create an emotional response that customers
have been tweeting about for years.

The illustration of Freddie's hand offering a high five created
a surprise moment that felt right to us, but we wanted to push
it further. We spent the better part of a month animating Freddie's arm and making it interactive so you could actually high
five him (FIG 5.2).

I remember thinking we were going way too far down the
rabbit hole at the time—but it ended up being worth the effort.
The Freddie high five launched resonated with people. Customers tweeted screenshots and videos of their hand hitting
their computer screen. One guy high-fived his iMac so hard,
he knocked it over and nearly broke it!

The response was so significant, we explored ways to expand
the high-five moment. T-shirts, a video game, and a rainbow
of foam Freddy hands later, what started as a peak moment
of emotional engagement became an extension of our brand
personality. (http://bkaprt.com/dfe2/05-03/).

The high five wasn't conceived as a marketing gimmick, but
after it spread like wildfire among Mailchimp customers, we

FIG 5.3: A high five can be even higher when you have a rainbow of Freddie hands.

learned a valuable lesson. Paying attention to peak experiences not only creates great experiences for customers; it's also good for business.

You can find your high-five moment, too. Take a careful look at the experience you're creating for customers and find the peak moments. Whether you use surprise as your strategy, or a more empathetic connection like that of the TurboTax example in Chapter 1, emotional engagement will be best received if you position it thoughtfully.

Now that we understand timing, let's look at some engagement methods that you can fit into any part of the customer journey.

DRAWING PEOPLE IN

Tell a story

For thousands of generations, humans have used storytelling to create intense emotional engagement with one another.

Stories can move us to tears, change our attitudes, opinions, and behaviors, inspire us—and change our brain chemistry.

A story that follows the traditional arc—with rising action, a climax, and a resolution—triggers two important neurotransmitters in our brains. The rising action of a story, which often includes conflict, causes our brains to release cortisol to focus our attention like a spotlight on concerning matters. Simultaneously, our brains release oxytocin, the neurochemical responsible for care, connection, empathy, and coincidentally, narrative transportation. With our attention focused and our minds thinking empathically, we're able to process the emotional experiences of others as if they were our own.

Two things result from the emotional transportation of a good story.

1. We process and recall information more effectively

In a 1969 study, Stanford professors Gordon Bower and Michal Clark discovered students were able to memorize and recall six to seven times more words when they organized the words into a story compared to when they made a random list (http:// bkaprt.com/dfe2/05-04/, PDF). Because stories shape emotions, and emotions in turn form long-term memory, stories are effective tools to help us understand and recall information.

2. Stories influence our actions

Neuroeconomist Paul Zak and his research team conducted a study on the effects powerful stories have on us. Researchers showed participants a short film about a two-year-old boy who doesn't know he will die from a growing brain tumor in a few months and the emotional turmoil of his father who knows the truth.

Afterward, the research team gave the subjects of the study opportunities to donate some of the money they earned for participating in the study to a charity related to cancer. Nearly all of the subjects donated, and on average, gave half their earnings. Zak and the team were able to predict with 80 percent accuracy

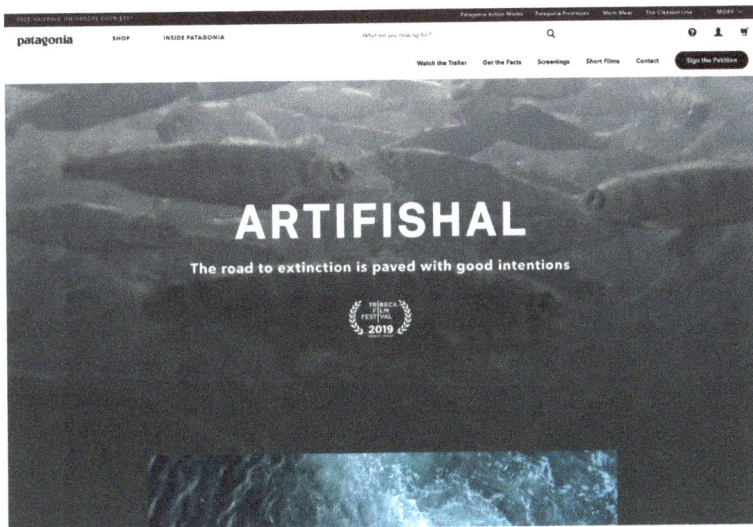

FIG 5.4: *Artifishal* tells the story of the perils of losing wild salmon populations. Nothing about the film sells their products, but it resonates with their eco-conscious audience (http://bkaprt.com/dfe2/05-07/).

who would donate by measuring the amount of oxytocin in each subject's blood (http://bkaprt.com/dfe2/05-05/).

Since stories cause the release of oxytocin, and oxytocin greatly influences us to take action, stories are an effective tool to call people to a cause.

Outside of a research environment, it may not always be obvious when you're being drawn in by a story. But even subtle story placement can be quite effective at creating a genuine sense of connection. For example, you may know Patagonia as the company that makes quality clothing and gear for outdoor experiences, but that's selling them short. They are a storytelling company.

Stories are fundamental to their mission and their products, promoting not just outdoor adventures for self-fulfillment but a broader sense of responsibility and stewardship for our planet. They produce documentary films to tell stories about threats to our wild spaces and wildlife, like *Artifishal* (FIG 5.4), which explores our disturbing influence on fish populations. They produce captivating content about the environment on their blog, *The Cleanest Line* (http://bkaprt.com/dfe2/05-06/).

Films and articles are expensive to produce, yet none of these stories is about selling their products. Patagonia tells stories about their hopes for the world, building a strong connection with like-minded people whose loyalty to the company is invaluable.

Each of the stories Patagonia tells builds respect for their brand, creates emotional engagement, and drives people to take action.

Embrace transparency

In a time when companies obfuscate their terms and conditions in pages of legal jargon, and many companies hide their more questionable practices, it's rare to find moments when a company is transparent about its struggles and shortcomings. When one does, the refreshing honesty can make a lasting impression that sets a business apart.

Stripe, whose business is online payment processing, has taken an admirably transparent approach to how they talk about their company and court talent. Examples of clear, direct copywriting and straightforward explanations abound throughout the site, and the Jobs page, in particular, includes an unusually candid discussion of what it's like to work for the company.

While most companies would go no further than a mission statement on this kind of page, Stripe continues on to share some of the problems they're confronting as a company, like a lack of people to staff great projects. They say something few tech companies admit: "We don't have all the answers." Most tech companies present themselves with hubris, as genius saviors who are making the world a better place. Stripe, by contrast, acknowledges their imperfections—and the vulnerability they show in this moment offers a natural point of emotional engagement that other tech companies competing for the same talent miss out on.

Perhaps the most compelling moment of transparency on this page comes further down, where they share the results of

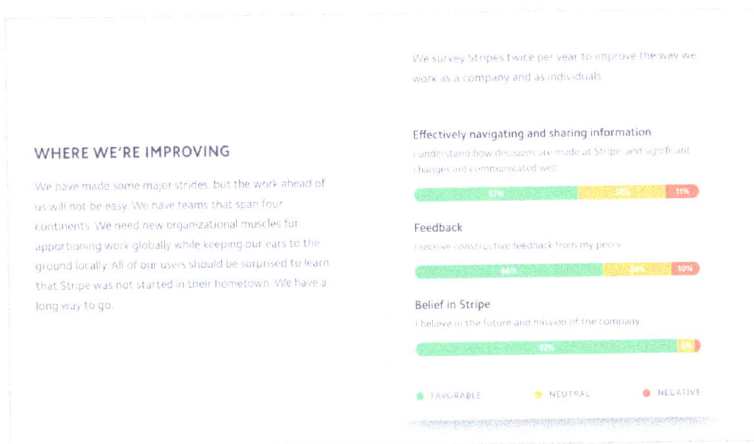

WHERE WE'RE IMPROVING

We have made some major strides, but the work ahead of us will not be easy. We have teams that span four continents. We need new organizational muscles for apportioning work globally while keeping our ears to the ground locally. All of our users should be surprised to learn that Stripe was not started in their hometown. We have a long way to go.

We survey Stripes twice per year to improve the way we work as a company and as individuals.

Effectively navigating and sharing information
I understand how decisions are made at Stripe, and significant changes are communicated well

Feedback
I receive constructive feedback from my peers

Belief in Stripe
I believe in the future and mission of the company

FAVORABLE   NEUTRAL   NEGATIVE

FIG 5.5: In a strong moment of transparency, Stripe publishes results of their biannual employee survey on their Jobs page, revealing the areas where they're doing well and where they still need improvement.

their biannual employee survey that shows obvious areas for improvement (**FIG 5.5**).

Why would Stripe show all this in the area of their site where they're trying to attract the best and brightest to their company? Simple: talented people interview with a lot of companies and recognize when something's being covered up. With radical transparency, Stripe shows candidates that they are honest with employees and foster an environment where everyone can grow together.

By designing for transparency, Stripe strengthens their recruiting efforts and creates a culture of accountability that influences product design decisions which influence customer perceptions as well.

Present a challenge (and constant rewards)

If you've ever used Duolingo to learn a language, you know that it's designed to be a bit addictive. Learning a new language requires memorization, lots of practice, and stamina to put in the time to achieve mastery. To help us succeed in our quest to

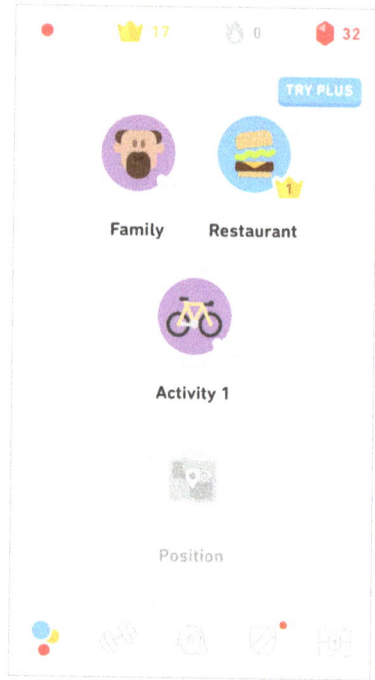

FIG 5.6: Duolingo locks portions of their language courses until a user achieves mastery of prior units. The unlock moments in their app trigger a dopamine release in our brains to encourage us to keep learning.

learn new languages, Duolingo offers constant positive feedback, rewards, and a method that might seem counterproductive to learning: locked areas of their app (**FIG 5.6**).

"Unlocking" moments give us a little hit of dopamine, the neurotransmitter that motivates us to take action toward goals, desires, and needs, and gives a surge of reinforcing pleasure once achieved. This builds a habit: the more we unlock, the more we want to push forward and do it again.

Duolingo want us to want to continue on our learning journey, so they lock portions until we've earned access. By breaking the learning process into discrete sections, they make an otherwise daunting task more approachable and challenge us to complete more units today than yesterday. With each unlock, we feel a sense of reinforcing pleasure that motivates us to keep learning.

THERE IS NO FORMULA

There are so many ways to design moments of emotional resonance into the user experience, regardless of your line of business.

We always want to be conscious of the timing of emotional engagement. Just as I'm not keen to accept hugs before coffee in the mornings, we can't expect to get positive reactions from our customers when we're overly clever during the low points of the customer journey. Mind the hierarchy of needs. Ensure you've addressed functionality, reliability, and usability before you layer on emotional engagement.

The peak-end rule reminds us that timing plays a major role in shaping the memory of experiences. Use it to your advantage to find your high-five moment as Mailchimp has.

The experiences you design can tell stories that form memories and drive action; they can unlock exciting new things, motivate with challenges, and they can offer moments of honesty that make people feel like they're talking with real human beings. No doubt you'll find new ways to build moments, too.

The examples we saw in this chapter use emotional design to engage an audience in the best of times, but what do we do when people harbor mistrust of our brand, are fearful, or when we've made a mistake? In the next chapter, we'll learn how to design for more difficult circumstances.

6 TRUST, FEAR, AND FORGIVENESS

THE PREVIOUS CHAPTER SHOWED us how we could use emotional design in many parts of the customer journey to draw people in, create memorable peak moments, and encourage deep engagement. Now let's look at how we can also use it to address substantial emotional obstacles like fear, skepticism, and mistrust—any of which can break a business if not addressed.

As we discovered in Chapter 2, our brains break up complex situations into simple concepts so we can evaluate the costs and benefits of a decision. To protect ourselves from harm, we're preprogrammed to be skeptical of new brands, products, situations, and even people.

Walk onto a used car lot, and your spidey-sense will tingle when the salesman approaches you with a preplanned pitch. We can smell bullshit a mile away.

That's what you're up against when you try to convince your audience to click, sign up, or trust your brand. It's you versus your audience's gut. You'll need to be persuasive without letting your marketing show when courting a skeptical audience. Before we can learn to overcome these obstacles, we need to dissect and examine the decision-making process.

GOING WITH YOUR GUT

We like to think that as the most highly evolved species to walk the planet, we navigate life with careful logic untainted by the baggage of emotion. We aspire to be more Spock than Kirk.

It's a noble idea, but far from the truth. In reality, we rarely have time to employ complex reasoning to make decisions, so we rely on gut reaction instead. Recall the decisions you made today, and you'll see that your gut is in the driver's seat.

What shirt should I wear? Hmmm, the blue one looks nice. What should I have for breakfast? Eggs and bacon sound great (probably not the best choice for me, but...). Crap, looks like there's traffic ahead. Maybe I'll take this exit to see if I can get around it.

Intuition drives so many decisions we make each day. You're wearing the shirt you have on now because you "just felt like it." You probably had other valid options, but if you used logic to consider each and every one, you'd never make it out the door. The problem is, often, there are several logical options to choose from, and logic can leave us gridlocked with no clear path to follow. Emotion is the tie-breaking vote when several options are equally valid. You use instinct to choose something that's good enough when the best option is unclear. If it weren't for gut decisions, we'd be lucky to get anything done.

So what would happen if emotion wasn't helping us to make decisions? Antonio Damasio, Professor of Neuroscience at the University of Southern California, has studied people who have injured the areas of the brain responsible for emotion. Basic decisions absolutely vex them. Deciding when to schedule a doctor's appointment triggers a circuitous internal debate of the various options. Similarly, choosing a restaurant for lunch proves impossible, as evaluating pros and cons never ends. Where there are numerous options of similar or equal merit, there's nothing to push these people's thought processes into a final decision. Without the tie-breaking vote the emotional gut response provides, they can't decide.

As designers, we're in a unique position to help users follow their gut instincts. Using common design tools like layout, color, line, typography, and contrast, we can help people more easily consume information and make a decision driven by instinct more than reason. Just as you chose the shirt you're wearing because it felt right, we can help our audience sign up for a service or complete a task because their gut tells them it's the right thing to do. Remember, we don't have to make an exhaustive case for action because reason is not often the primary driving force our audience uses to decide. We just have to appeal to their emotions to make the benefits appear to outweigh the costs.

As we'll soon see, there are so many ways to account for negative emotions as you design.

Wealthsimple: Trust in simplicity

Close your eyes. Come on, play along with me. Close your eyes and picture a bank website. Hold that image in your mind. How would you describe what you saw? Was it simple, clear, or interesting? I'd wager a shiny nickel that none of those words came to mind.

Instead, you probably saw something busy with generic photography of smiling people and a scrolling carousel up top shilling irrelevant products and services. The language is verbose, technical, and confusing. Sound right? Banks want their websites to feel personal and unintimidating, but the subtext rises to the surface. They're very confusing businesses.

I chatted with Rudy Adler, cofounder and chief product officer of Wealthsimple, and he shared with me that he thinks banks are confusing by design: "Banks are so distrusted because they have a very confusing way of talking about things and presenting fees. It's almost like their business model is purposely confusing."

Adler, who cut his design teeth in an experimental program at Wieden + Kennedy, considered the emotional experience of banking carefully with his cofounders when they started Wealthsimple. Their Webby-award-winning website, a platform for automated investment decisions, is anything but cluttered (FIG 6.1).

Adler told me, "We consider our design and brand a competitive advantage. We want to take a different approach. We want to walk the line between simple and smart. We want our design

---

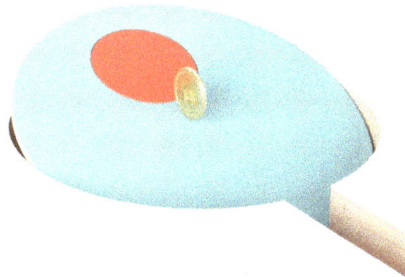

Investing on autopilot

FIG 6.1: Wealthsimple's website uses design as a competitive advantage. Impeccable execution and plain language walk the line between sophistication and simplicity.

to feel human and simple, but not so simple that the customer feels we're not sophisticated."

It comes through clearly on their homepage, which features 3D animations of enchanting Rube Goldberg machines. They appear so realistic that I naively asked Adler how they built and shot the machines, and he told me they worked with the best 3D animators in the business to produce what some might call non-essential parts of the page.

But those elements are essential to the message—they shape their customers' first impressions.

We see our homepage as the first glance in what we hope will become a relationship that lasts a lifetime. So that first impression is pretty crucial.

We know what we do is different—it's hard to explain auto-mated investing to someone who's never heard the phrase automated investing, let alone to someone who's never invested a penny. We wanted to make a site that provided information as simply, clearly, and beautifully as possible. And we wanted a central metaphor that was fun, elegant, and the opposite of tech-confusing.

So we went with a Rube Goldberg machine. Yes, yes, we know a Rube Goldberg machine is an overly complicated contraption to achieve a simple task. But that's part of the humor. Because we see what we do as something very simple designed to achieve a really complicated task. (http://bkaprt. com/dfe2/06-01/)

The care and consideration apparent in the design gives users the impression that equal attention is paid behind the scenes to the inner workings of Wealthsimple. Rather than endeavoring to describe the novel investing algorithms they've developed with financial jargon-riddled copy, they demonstrate their level of sophistication with a visual metaphor; showing, not telling. Their design shapes first impressions by inspiring trust.

"We're in financial services, so trust is pretty important," says Rudy Adler. That's a bit of an understatement; anyone who's managing your money had better be running a tight ship, right? At publication, Wealthsimple had converted 175,000 skeptics into believing customers, in no small part because of the power of emotional design.

Apple: Trust as competitive advantage

The phone market is brutally competitive. At this point, the hardware and software of mobile devices are all pretty comparable. Whatever phone you choose, you're going to be able to take great photos, manage essential communications, and stay entertained. Where do you find a competitive foothold if you're in a neck-and-neck race with competitors?

Apple is one of the few companies to recognize that amidst never-ending stories of data breaches and misuse plaguing their competitors, trust can be a competitive advantage—a clear emotional need of their customers that remains unmet by competitors. And they highlight this on their website (FIG 6.2).

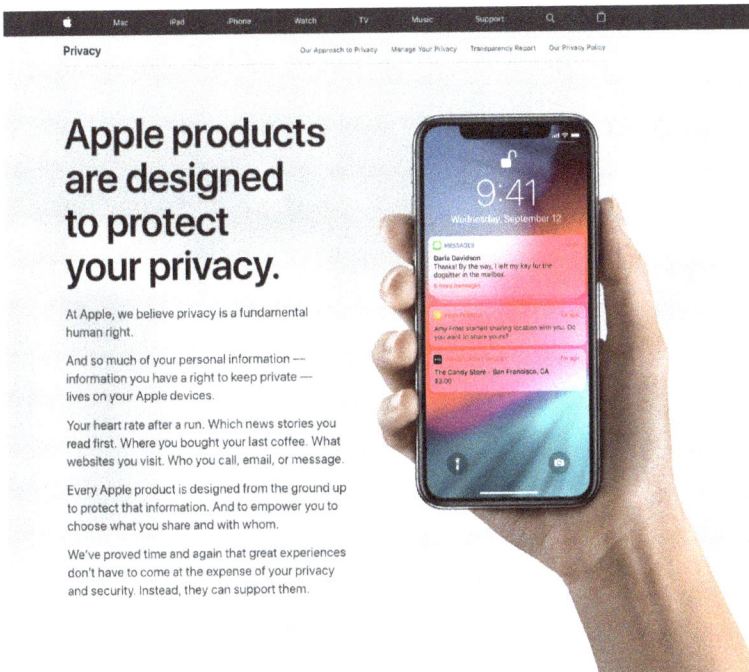

FIG 6.2: Privacy is treated as an important differentiating feature on Apple's website. In an era of privacy invasion perpetrated by external and internal governmental agencies, it's quickly becoming one of Apple's most compelling features.

Data privacy, device privacy, the rules third-party developers abide by, and their financial transaction practices are all spelled out in plain language for the layman to understand and consider.

Those considering a switch to a cheaper phone on another platform may pause and consider if trading their privacy for a little extra savings is worth it. Apple is tipping the scales in their favor by investing in privacy features and making them a central part of their marketing efforts. It's a tactic that creates a smart competitive advantage.

There are so many places in the customer journey where trust in your company can break down. Designing for all of those moments can be difficult. Airbnb has taken an interesting approach to designing for trust across teams and throughout the entire customer journey.

Airbnb: Mapping trust

In Chapter 4, we saw an example of a gap in Airbnb's intentions to create trust between hosts and guests and the impact that inadvertently enabled hosts to deny bookings based on race. They haven't always gotten things right, but none of us have. Mistakes are learning opportunities that can show us how to map a better way forward.

Airbnb did just that: they mapped the customer journey to identify where trust is most likely to break down.

At the heart of their business model is the often daunting idea of inviting strangers to stay in your home. As design manager Adam Glynn-Finnegan of Airbnb told me,

> The only thing that would make that work is establishing trust. Everything we do at Airbnb is possible because of trust—the trust people put in each other and in us. It's the core innovation of our platform, and it's also the currency that circulates throughout the entire Airbnb experience, bringing together and bonding guests and hosts.

Airbnb sees their role as the facilitator of trust between hosts and guests, and in that role they have to ensure everyone's safety. Like Apple, they explain their systems and processes in plain language to customers to earn their trust (FIG 6.3).

Trust is so important to the company that they've built a Trust team, which carefully examines the customer journey for points where things can go wrong, and proactively develops preventative measures.

The method they use to do this will sound familiar: Airbnb uses the same concept of the customer journey map we explored in Chapter 4 to create what they call a "Trust Map."

The Trust Map helps visualize and deconstruct the spectrum of moments when trust can be built—or eroded—for both guests and hosts. By creating this shared understanding, it

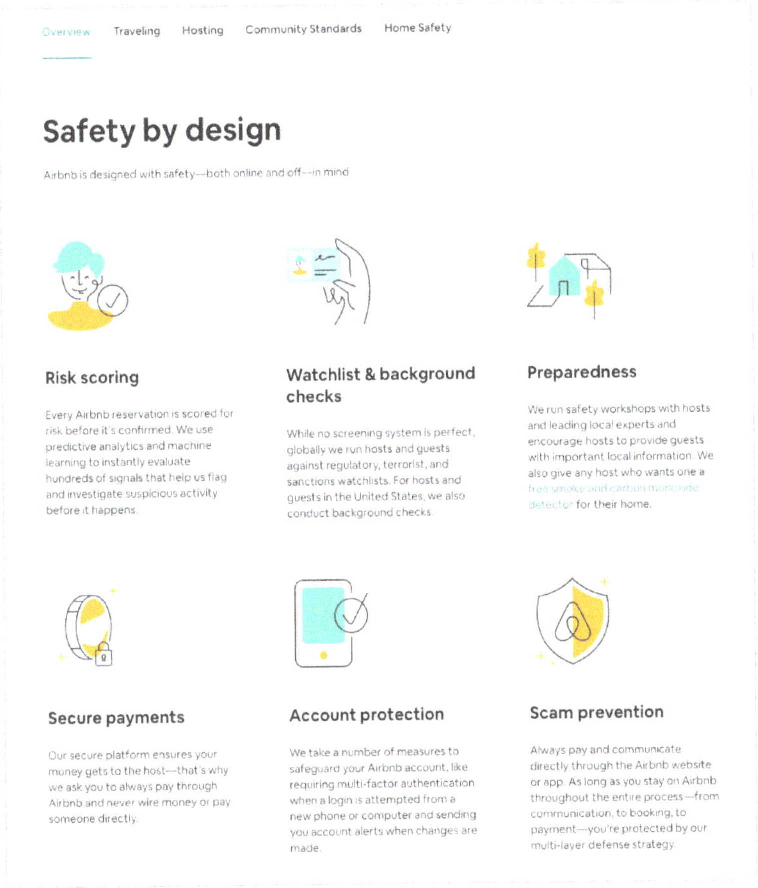

Overview
Traveling
Hosting
Community Standards
Home Safety
Safety by design
Airbnb is designed with safety—both online and off—in mind
Risk scoring
Every Airbnb reservation is scored for risk before it's confirmed. We use predictive analytics and machine learning to instantly evaluate hundreds of signals that help us flag and investigate suspicious activity before it happens.
Watchlist & background checks
While no screening system is perfect, globally we run hosts and guests against regulatory, terrorist, and sanctions watchlists. For hosts and guests in the United States, we also conduct background checks.
Preparedness
We run safety workshops with hosts and leading local experts and encourage hosts to provide guests with important local information. We also give any host who wants one a free smoke and carbon monoxide detector for their home.
Secure payments
Our secure platform ensures your money gets to the host—that's why we ask you to always pay through Airbnb and never wire money or pay someone directly.
Account protection
We take a number of measures to safeguard your Airbnb account, like requiring multi-factor authentication when a login is attempted from a new phone or computer and sending you account alerts when changes are made.
Scam prevention
Always pay and communicate directly through the Airbnb website or app. As long as you stay on Airbnb throughout the entire process—from communication, to booking, to payment—you're protected by our multi-layer defense strategy.

FIG 6.3: Airbnb pulls back the curtain to explain how they ensure the safety of hosts and guests to inspire trust in their business (http://bkaprt.com/dfe2/06-02/).

uncovers opportunities to create a community where everyone feels safe and secure (**FIG 6.4**).

With this tool at hand, trust is a core focus no matter what part of the Airbnb platform a team may be working on.

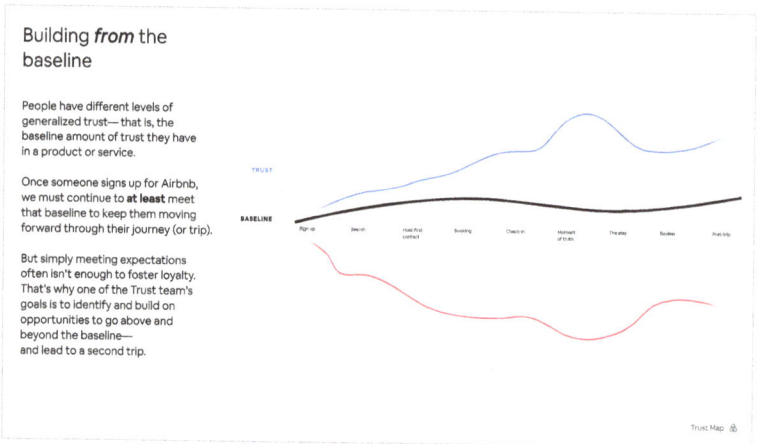

Building *from* the baseline

People have different levels of generalized trust— that is, the baseline amount of trust they have in a product or service.

Once someone signs up for Airbnb, we must continue to **at least** meet that baseline to keep them moving forward through their journey (or trip).

But simply meeting expectations often isn't enough to foster loyalty. That's why one of the Trust team's goals is to identify and build on opportunities to go above and beyond the baseline— and lead to a second trip.

TRUST

BASELINE

Sign up   Search   Host First   Booking   Check In   Moment   The stay   Review   Post-trip
                   Contact   of trust

Trust Map

FIG 6.4: Airbnb's Trust Map plots out key moments in the customer journey where trust can be created or eroded, so teams across the company can contribute to the common goal of creating amazing experiences for their customers.

ACKNOWLEDGING FEAR

Part of designing for trust is simply identifying and addressing fears head-on. You'll see the principles of authenticity and transparency from the last chapter carry forward here, but now the stakes for customers are a bit higher.

Nothing is more foundational to our personal identity than our genetic code. It determines our physical form, influences potential health issues, and plays a role in shaping behaviors and preferences.

Locked inside our DNA are the traits of our ancestors, clues that help us uncover their stories, and traces of their legacy that live on in us today. It's tantalizing. Who *wouldn't* want to unlock these powerful clues of our past, present and future?

Despite a strong value proposition, many people have serious fears about hiring a DNA sequencing service. Members of my family resisted my invitations to join the service for fear it would impact their health insurance and pose privacy issues. 23andMe neutralizes these fears and others on their website (**FIG 6.5**).

23andMe   OUR SERVICES ∨   HOW IT WORKS ∨   REPORTS   STORIES   SHOP   SIGN IN   REGISTER KIT   HELP ∨

It's just saliva.
No blood. No needles.

Our home-based saliva collection kit is all you need to
send your DNA to the lab. We have made the process as
simple as possible.

shop now

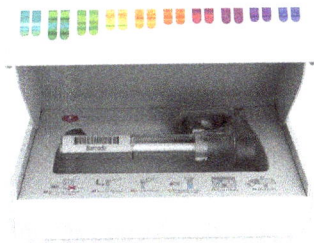

Three steps. It's simple.

All from home. No blood. No needles. Just a small saliva sample.

1 Order

Choose from our Health +
Ancestry or Ancestry + Traits
services. Your saliva collection kit
is the same for both services and
typically arrives within 3 to 5
days. Express shipping is
available.

2 Spit

Follow kit instructions to spit in
the tube provided – all from
home. Register your saliva
collection tube using the
barcode so we know it belongs
to you, and mail it back to our lab
in the pre-paid package.

3 Discover

In approximately 3-5 weeks, we
will send you an email to let you
know your reports are ready in
your online account. Log in and
start discovering what your DNA
says about you.

FIG 6.5: 23andMe addresses the fears of potential customers straight away on their homepage to improve conversion rates.

First, they tell us, "It's just saliva. No blood. No needles." Whew! What a relief.

Next, they establish their bona fides by explaining the scientific rigor behind their testing, their FDA approval, and their various third-party certifications.

Then they address the issue of privacy. They tell us that our genetic information is protected by federal law and cannot be shared with insurance companies, employers, or anyone else without our express consent.

Finally, they share some technical information about how our genetic information is decoupled from personally identifiable information to protect our identity.

Most marketing websites focus all efforts on telling potential customers how amazing they are, but 23andMe knows their audience well. They recognize that fear is top of mind for their visitors. By addressing fears directly, they shift the buyer's emotional focus from fear to curiosity.

Staying calm in crisis

Our instinct to take action in a crisis is a challenge that financial companies like Wealthsimple must consider. When the economy takes a bad turn, fear can drive investors to make really bad decisions. Our instincts tell us to run from a bad situation, but the worst thing you can do when the stock market tanks, is sell your holdings. It's better to sit tight and ride it out knowing that in the long term the markets will rise again. Let's take another look at Wealthsimple to see how they keep customers from making bad financial decisions when things are looking bleak.

Unlike most companies, Wealthsimple doesn't want to drive their customers into their products daily where they might be shaken by sudden market downturns.

"Emotion is the enemy of smart investing," observes co-founder Rudy Adler. "We want to optimize for infrequent use to minimize the negative emotions that come from watching your portfolio go up and down with the markets."

When customers use the portfolio app, the charts they see by default show long periods of time where the upward trend is visible. Showing short time scales of a month or a week give the impression the sky is falling, but the zoomed-out picture gives cause for optimism. This design brings contributions to the forefront, shifting emotion from fear to confidence (FIG 6.6).

Thoughtful design that addresses the emotions of their customers have helped Wealthsimple inspire trust, diminish fears, and shape their customers' smart investing behaviors.

Designing for trust and mitigating negative emotions like fear help us build stronger relationships with our customers, but what do we do when we inevitably make a mistake? Emotional design can also help us right our wrongs and make forgiveness possible.

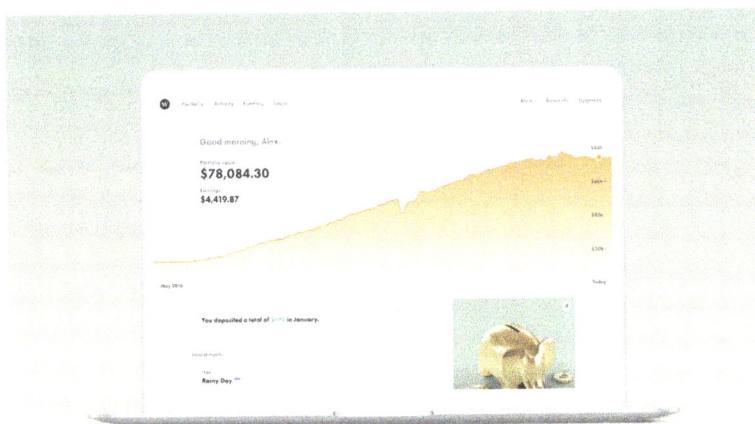

FIG 6.6: Wealthsimple's portfolio app shows a broad timescale of performance, yearly rather than monthly or weekly, to diminish fears that come with market volatility and lead to bad investing behaviors.

FORGIVENESS

Sooner or later, something will go wrong with your website or product. Servers go down, people make mistakes, and the unforeseeable happens. In such situations, it's helpful to have your audience's goodwill on your side so they will more easily overlook a temporary shortcoming and maintain trust in your brand.

Flickr knows from firsthand experience that a good response to a bad situation is critical. It doesn't hurt to have a devoted fan base too, as we'll see.

Flickr: Turning lemons into lemonade

A long time ago in a galaxy far, far away, before Instagram and Snapchat, the photo sharing platform Flickr reigned supreme. It was much beloved for its simplicity and function, but also for the thoughtful emotional design that was carefully tucked into all the right places in their site. To this day, Flickr still occupies a warm place in the hearts of many.

But in July of 2006, a massive storage failure took Flickr and millions of people's photos offline. Though all photos were safe, and no data was lost, thousands of avid users worried as their favorite photo site took a temporary nap (roughly three hours). Tensions ran high as engineers worked to bring the site back online. Inquiries from concerned customers poured in.

During the crisis, the Flickr team had a stroke of genius. Thinking like a veteran parent trying to keep an antsy kid occupied while waiting for food in a restaurant, they applied the art of redirection and ran an art contest. They posted a message that explained the outage and asked users to print the page and do something creative with it to win a free, one-year Flickr Pro account (FIG 6.7).

Rather than brooding over their missing photo library, users brainstormed ways to win the prize. Hundreds of entries were submitted—some of which were very clever (FIG 6.8).

Though the site was down and many were inconvenienced, Flickr users remembered the fun they had participating in the coloring contest, and for some, how great it was to win a free year of Pro service.

Flickr survived their crisis, and in doing so left us lessons we can carry forward into our work today. It's important to confront the negative emotions that arise in situations like this, and the experience you've designed around your site just might save you.

Flickr worked through the stressful situation by communicating calmly and honestly with their users. Let's take a closer look at how Flickr handled the event to learn how emotional design shaped user reactions.

Lesson 1: Address concerns clearly and head on

During events like the one Flickr experienced, the right tone is essential to easing concerns. When people are deeply stressed by an outage or a mistake you've made, you must explain what happened swiftly, honestly, and clearly. Give people the facts of the event, communicate that you're doing your best to resolve

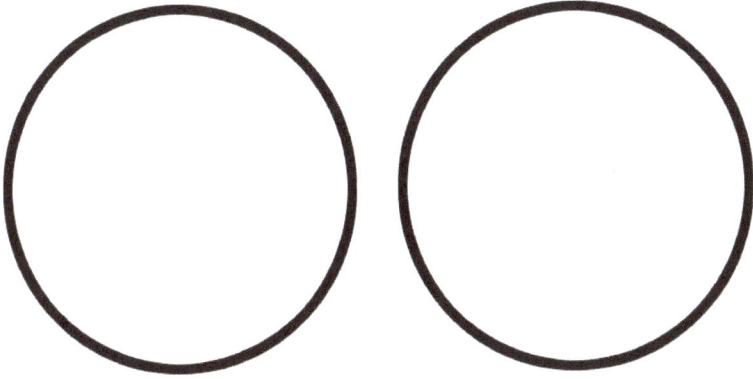

flickr

Arrggh! Our tubes are clogged!

Because this sucks˙, we thought you might like to enter an impromptu competition to win a FREE PRO ACCOUNT!

Just print out this page and colour in the dots. When the site's back up, take a photo of your creation and post it to Flickr, tagged with "flickrcolourcontest".

Team Flickr will pick a winner in the next couple of days, and that lucky duck will get a free year of Pro.

˙ Seriously, we apologise for the unannounced downtime. We're working as fast as we can to get flickr.com back online. Details here.

FIG 6.7: During a major outage in July of 2006, Flickr ran an art contest that turned stressed users into content contestants.

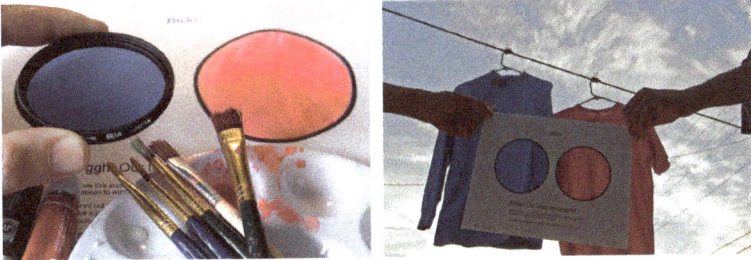

FIG 6.8: People went into creative hyperdrive over the Flickr art contest, submitting clever entries that won a select few a free Pro account. Photos by KC Soon (http://bkaprt.com/dfe2/06-03/, left) and Bart Kung (http://bkaprt.com/dfe2/06-04/, right).

things quickly, and update users regularly, even when little has changed. That's exactly what Flickr did via their blog as the event unfolded (http://bkaprt.com/dfe2/06-05/).

Updates let people know you're still focusing all of your attention on resolving the problem. They give you another opportunity to apologize for the inconvenience and reassure your users that you'll fix the problem as quickly as possible.

Once you've done your best to soften emotions, you might consider a redirection like Flickr's. Giving users something for free can rekindle the goodwill you've worked so hard to cultivate and gives them something else to focus on while you do your best to fix the problem. If giving something to everyone isn't possible, a contest is a nice way to achieve the same redirection effect while limiting the expenses you may incur.

In high-stress situations, your top priority must be to tame negative emotions as best you can and, if possible, shift them back to the positive.

Although their clever response to the outage helped save the day for Flickr, it wasn't the only reason their users stuck with them in a time of crisis.

Lesson 2: Invest in emotional design today to reap benefits tomorrow

What really saved Flickr on July 19, 2006, wasn't just a clever contest; it was the emotional design in their website that accrued user devotion. Flickr was one of the first websites to understand the power of emotional design. They created an informal and human personality in the product that made it a joy to use. The art contest was simply another way for the design persona that earned them a devoted following to manifest itself. Sure, people get upset when they can't access one of their favorite apps, but a long history of great experiences with the product outshines the inconvenience of an outage.

Emotional engagement before and during a major event can help mitigate the risk of losing your audience. As we saw earlier in this chapter, your audience performs an internal cost-benefit analysis every time you ask them to complete a task. When something goes awry and your audience is inconvenienced, there's a risk that users will suddenly perceive the costs of using your product as greater than the benefits.

Think of emotional design as an insurance policy that can help maintain audience trust when things aren't going your way. Emotional engagement can help us look past even the most serious infractions, leaving the good more indelible in our mind than the bad. Psychologists call this phenomenon of positive recollection the *rosy effect*. As time passes, memories of inconveniences and transgressions fade, leaving only positive memories to shape our perceptions.

This is good news for designers, as it means that the inevitable imperfections in our work don't necessarily lead to mass user exodus. Donald Norman, author of *The Design of Everyday Things* and director of the design lab at UC San Diego, points out that pursuing perfection is a spurious goal, as the total experience we're creating will shape our users' memories of our work in the end (http://bkaprt.com/dfe2/06-06/):

> As interaction designers, we strive to eliminate confusion, difficulty, and above all, bad experiences. But you know what? Life is filled with bad experiences. Not only do we survive them, but in our remembrance of events, we often minimize the bad and amplify the good.
>
> We should not be devoting all of our time to provide a perfect experience. Why not? Well, perfection is seldom possible. More importantly, perfection is seldom worth the effort. So what if people have some problems with an application, a website, a product, or a service? What matters is the total experience. Furthermore, the actual experience is not as important as the way it is remembered.

Though carefully and considerately responding to mistakes and problems will help get you out of hot water, the emotional design groundwork you lay before an event will keep your audience committed to your brand. The forgiveness we earn through careful emotional design can prevent considerable losses in customers and revenues, which is alone a compelling enough reason to incorporate it into our design process.

We're approaching the end of our time together as we jump into our final chapter, and I'll admit, I'm feeling a bit sad about that. We've come so far together!

In the last chapter, we'll talk about how you can help your colleagues see where emotional design fits into our process even when you're trying to create a minimum viable product. We'll also see some examples of companies that have won big by employing many of the ideas we've learned in this book.

Let's go!

THE BUSINESS OF EMOTIONAL DESIGN

NOW THAT YOU'VE LEARNED the principles of emotional design and seen them applied with many familiar brands to great effect, you may be excited to put these methodologies into practice in your own work. But some of your colleagues may push back, perhaps unsure how it could be folded into your process, or unaware of the value emotional design brings to the user experience (and ultimately to companies).

In this chapter, I'll share a model to guide how emotional design can fit into a fast-paced release schedule. I'll also share some examples of emotional design that have created quantitative business outcomes like market adoption rate, company valuation, and capital invested. Having examples on hand will help you start conversations in your company about the value of emotional design.

You know how emotional design works. Now let's bring your colleagues on board so you can put it into practice together.

BRINGING EMOTIONAL DESIGN INTO YOUR PROCESS

It's better to build something that a small number of users love, than a large number of users like.
—SAM ALTMAN, Y COMBINATOR CHAIRMAN
(http://bkaprt.com/dfe2/07-01/)

I've worked in and led many types of teams in agencies, in product companies, and in marketing organizations. Though the type of work varied, one thing didn't: there's always a sense of urgency to produce and ship as quickly as possible.

Speed is sacred, especially in teams following the popular Agile methodology for software development (http://bkaprt.com/dfe2/07-02/). In software, we optimize for speed so we can find product-market fit. Will our customers like this? Does this solve their problems?

If Agile is the law of the land in your company, as you talk to your colleagues about bringing emotional design into your process you may hear, "That sounds nice, but we can add that after we build our MVP." Sound familiar?

An MVP is the "minimum viable product," or the simplest thing you could make that would let you get to market quickly to test your product with real customers.

Engineers and executives have promised me a hundred times that we'll return to our emotional design ideas after we ship our MVP, but it almost never happens, as urgent tasks surface in each new iteration. This happens because there's a flaw in our thinking about MVPs and what they should include.

In Chapter 1, I shared a model of the hierarchy of user needs. Our customers need functional, reliable, and usable products, but to create extraordinary experiences that create sustained business value—like those you'll see later in this chapter—we need to also design for emotional engagement.

When creating an MVP of a site or product, the common approach is to focus on the bottom three layers of the pyramid: functionality, reliability, and usability. (In many cases, usability isn't even a high priority in MVPs). Rushing to market and

Minimum Viable Product

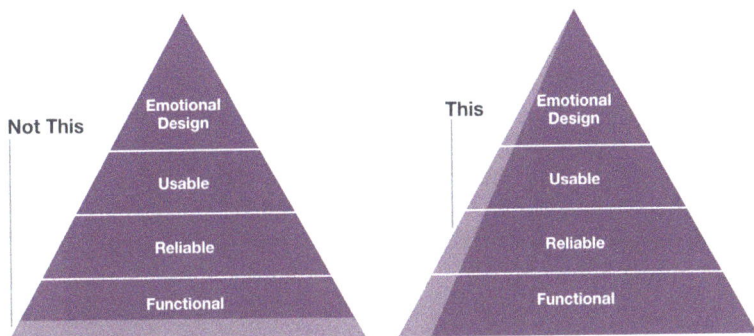

FIG 7.1: MVPs should be a vertical slice of the hierarchy of user needs, not limited to basic functionality (Diagram by Jussi Pasanen, http://bkaprt.com/dfe2/07-03/).

coming back later to add emotional engagement is a flawed approach, as it squanders the first impressions your customers have of your product. If their experience isn't great the first time, they're not coming back.

When we download an app to our phone and use it for the first time, we'll delete it if the experience is poor. It's rare to download the same app again for a second try. The same is true of almost anything on the web. If you don't nail your first impression, you're going to see very high churn rates.

That's why emotional design shouldn't be treated as a layer of the experience to add *after* you've gone to market. It should be integrated in manageable pieces with every iteration.

After reading the first edition of this book, Australian designer Jussi Pasanen created a clever modification of the hierarchy of needs diagram to help us reframe the way we think about MVPs and emotional design in an Agile process (FIG 7.1).

Each layer of the hierarchy—functionality, reliability, usability, and emotional design—should be included in every iteration of a product, even the MVP. With this approach we can still go to market quickly, but with products that people will actually want to use.

So how might we design for emotion in a sprint to build an MVP? Here's an uncomplicated process that uses the tools from this book and won't compromise your speed.

- **Step one: Consider the emotional needs of the customer.** Are they first encountering your product with hesitations that could prevent their purchase, as we saw with 23andMe? Is inspiring trust important, as was true with Wealthsimple? Be scrappy and talk to a handful of potential users, then make an empathy map (see Chapter 4) to understand the emotions you're designing for.
- **Step two: Map the customer journey**—the steps they'll go through to use your product. Consider where they might experience friction and where they might have a peak moment. Rapid prototyping and testing with users will help you see the customer journey more clearly and inform what you'll build.
- **Step three: Design one peak moment in your MVP** in the most critical part of the customer journey. Remember the peak-end rule (Chapter 3)—your moment will have maximum power if positioned at the end of an important step in the customer journey.
- **Step four: Consider how you might measure the impact of your peak moment.** Metrics could include mentions on social platforms, customer support tickets, return sessions, time in product, customer retention, or task completion. Find the right metrics for your situation and monitor them carefully after launch. This data can help you illustrate to colleagues the role emotional design plays in your strategy.

It's as straightforward as that. If you have the time to create a design persona in the early stages of your product, great, but don't let perfect be the enemy of good. Designing peak moments are a great place to start.

These steps are lean enough to fit into any Agile sprint and shouldn't bloat your design process. If you and your team are in the habit of running quick design sprints, you'll find it easy to fold these steps into your process. They're not a huge time investment, either, so if you're the only designer on a cross-functional team, you can certainly go through each of these steps on your own.

Don't ask anyone for permission. Act like the expert you are and follow the process you know is essential to delivering the best design.

Step four of the above process is worth further mention because part of acting like a professional is knowing how your work helps your business succeed. Communicate regularly with colleagues about the value your work creates. If you can learn to think and speak about design not just in qualitative language ("the five customers we interviewed found our onboarding process much easier to follow"), but in terms of business outcomes ("we improved customer adoption by 22 percent"), your career trajectory will dramatically improve.

Knowing how emotional design has created real value for other companies is also helpful. Having a few case studies to reference in conversations with skeptics will help you bridge emotional design to something all your colleagues can understand: the success of a business.

SPEAKING TO THE BUSINESS VALUE OF EMOTIONAL DESIGN

We designers often make decisions based on our strong value system. We strive to design with a user-centric approach. We want to create things that are both usable and beautiful. We want our designs to work as a system across screens, devices, and channels to create an elegant, consistent user experience.

These values are largely qualitative. They're squishy, hard to measure, but important. Our colleagues in engineering, product, marketing, sales, and executives with whom we partner daily think differently. Their values tend to be more quantitative and measurable. They want to track things like time to market, churn, adoption rate, company valuation, and Net Promoter Score.

These two ways of thinking aren't contradictory but complementary. As such, whether the value system that guides our thinking is largely qualitative or quantitative, we're better off if we're able to speak about our work from both perspectives.

We've already seen some compelling stories of how emotional design played a major role in big business outcomes. Let's revisit a few, but through the lens of business goals.

Category creation: Emotional design helped Headspace do something extremely difficult. They successfully defined a new meditation and mindfulness product category and attracted millions of users.

Customer acquisition and retention: Wealthsimple acquired customers by using emotional design to inspire trust in their platform, and they retain customers by mitigating fears that could cause investors to sell their holdings and close their accounts.

Session length and return rate: Duolingo created a sticky learning experience that uses rewards and unlock moments to increase the length of learning sessions and brings users back for new sessions.

Growth and market reach: Mailchimp differentiated itself in a crowded market by building a brand with a personality that attracts 14,000 customers to sign up every day (http://bkaprt. com/dfe2/07-04/). Their high-five moment inspired countless customers to post on social media about their love for the brand.

Now that we're starting to see the connections between emotional design and quantifiable business metrics, let's look at two more case studies. The widespread success seen by messaging platform Slack and calendar app Sunrise are both thanks to their foundational investments in emotional design.

Slack: Stealing a market

Andrew Wilkinson, CEO of the design agency MetaLab, groaned as he read an email from Stewart Butterfield in July of 2013. Butterfield, cofounder of Flickr, was pivoting his video game company Glitch to create a new messaging product called Slack, and he wanted to hire MetaLab to design it.

Wilkinson knew the messaging app space was crowded and felt it was a problem that didn't need solving yet again. Despite his skepticism around the product, he was a fan of Butterfield's work and decided to take on the project anyway. Little did he know, it would go on to become their most successful project of all time (http://bkaprt.com/dfe2/07-05/).

Slack was already live by then and had regular users, but it was visually basic. Wilkinson described it as, "IRC in the browser. Barebones and stark."

The MetaLab team helped Slack build a personality into the product that would differentiate it from competitors, all of which felt largely corporate and conservative. Wilkinson described competitors as HAL9000 from *2001: A Space Odyssey*—a stiff approximation of humanity with rigid language. Slack, by contrast, is like TARS or CASE from *Interstellar*, quick-witted and, as cowriter of the movie Jonathan Nolan describes, even more human than the people they serve (http://bkaprt.com/dfe2/07-06/, video).

By defining the personality, MetaLab was able to make design decisions that would bring life to the product and shape the emotional engagement that helped fuel its success. Wilkinson describes it best:

> With Slack, a bubbly, bright UI, delightful interactions, and hilarious copywriting come together to create a personality. A personality which has triggered something powerful in its users: they care about it. They want to share it with others. It feels like a favorite coworker, not a tool or utility.
>
> When you hear people talk about Slack they often say it's "fun." Using it doesn't feel like work. It feels like slacking off, even when you're using it to get stuff done. But when you look under the hood, it's almost identical to every other chat app. You can create a room, add people, share files, and chat as a group or direct message one another. (http://bkaprt.com/dfe2/07-05/)

Those of us who used Slack early on and witnessed its growth will recall that it lived on the lips of influencers and was recommended by many. It made people feel cool to use it. It felt fun. Feelings like that are powerful—they drive adoption.

MetaLab delivered a design that was successfully differentiated from the competition. Upon the release of the newly

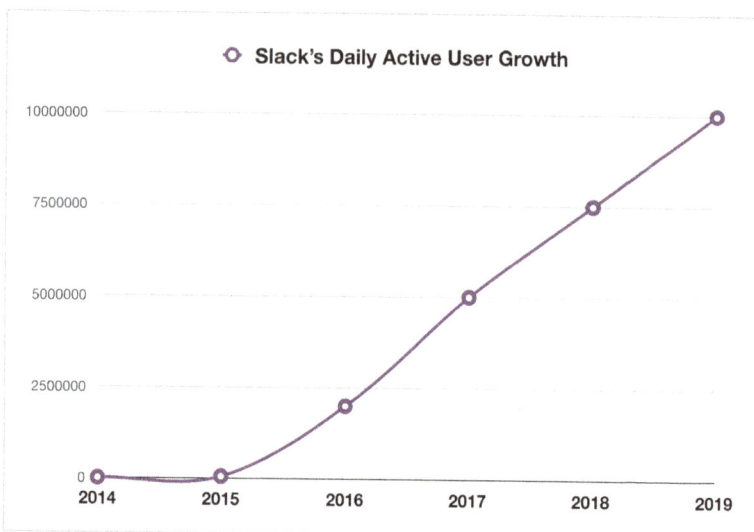

FIG 7.2: Slack's daily active user numbers grew rapidly after the release of the newly designed product, despite the fact that a number of strong competitors had already claimed a big piece of the market.

designed product in 2015, the growth of Slack's daily active users (DAUs) was meteoric (**FIG 7.2**).

In January 2019, Slack hit a major milestone: 10 million daily active users and 85 thousand paying customers. In April of the same year, Slack went public on the New York Stock Exchange with a market cap of roughly $13 billion. Not bad for a failed video game company that pivoted into an already crowded market.

There's more to Slack's incredible story. Atlassian saw Slack stealing the market from their product HipChat—which had been on the market since April of 2009 and was the leader in the workplace chat product space. They continued to iterate on the product, but in 2017, they introduced its replacement called Stride, which looked an awful lot like Slack (http://bkaprt.com/dfe2/07-07/). They finally realized that HipChat didn't have the appeal it needed to be competitive, and with Stride, they could make a fresh start with emotional design central to the product.

Unfortunately for Atlassian, it was too little too late. By then, Slack had captured a large part of the market and had a great deal of momentum. In July of 2018, Atlassian announced that they'd be sunsetting HipChat and Stride and were entering into a strategic partnership with Slack (http://bkaprt.com/dfe2/07-08/). It was official. Slack had won.

Designing for emotion created real value for the company—billions of dollars of it—and helped drive unprecedented growth of monthly active users.

Here's the moral of this story: design is rarely seen as a tool for creating durable competitive advantage, but that's exactly what it was for Slack. Product design and clever use of personality differentiated their product, changing the trajectory of the company and the entire business chat app market.

Sunrise: Surviving and Succeeding with Emotion

The grueling personal investment required to be an entrepreneur is hard, but harder still is the process of convincing investors to make a capital investment, especially when competitors have long since laid claim to the market.

The team behind Sunrise, Jeremy Le Van and Pierre Valade, believed that the poorly designed calendaring apps native to our phones were ripe for disruption. They applied for a spot in Y Combinator (YC), the seed accelerator that's helped launch over 2,000 companies, including Stripe, Airbnb, Instacart, and Dropbox.

Sunrise was entering an extremely crowded space, and they were competing with apps that were installed on every phone by default. Would thousands of people abandon the calendar integrated into their phone for a new product? Le Van and Valade thought so, but they were rejected from the program as investors were skeptical Sunrise could upend the status quo.

What those investors didn't see was the unique approach that Sunrise would take to bring users to a new calendar platform. Color, a generous layout, elegant interactions, and profile photos that put faces to names all brought life, warmth, and personality to the product. Other calendar apps felt like cold utilities by comparison (FIG 7.3).

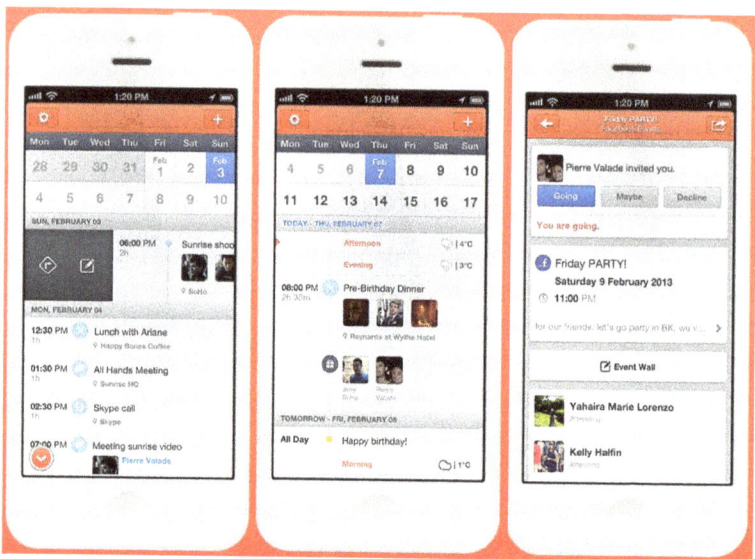

FIG 7.3: Sunrise pulled profile photos from Facebook and Google to create a calendar experience that was uniquely warm.

Despite being turned down by YC, Le Van and Valade built and launched Sunrise and began to build a strong user base with the warm, personable calendaring experience they knew was lacking from other apps. The cost to run the infrastructure was difficult for Le Van and Valade to carry without investment, and the product might have been shuttered if it weren't for investors who recognized how Sunrise was delighting users unlike any competitor.

Mike Hirshland, cofounder of Resolute Ventures, offered some insight after making a seed-stage investment in Sunrise:

The biggest Consumer Internet successes typically have offered little or no real technology innovation. Rather, they addressed a real consumer problem with a product that users loved, grew rapidly, and ultimately built barriers [to competition] either by deeply entrenching themselves in a user's daily life, and/ or by creating network effects. Meanwhile, competitors with more interesting technical innovation but less compelling user experiences failed to get adoption.

The lesson for me is this: the first challenge for consumer facing businesses is to build an awesome product that users love. This is table stakes to winning the market, and, for me the sine qua non to making seed stage consumer investments. As a consequence, at Resolute we typically look for awesome product/user experience founders, and we often pick teams focused on delighting users over teams with more impressive or ambitious technical innovation.

Only time will tell, but we think their intense focus on delighting users is going to win. (http://bkaprt.com/dfe2/07-09/)

What Hirshland wrote could be said of most of the examples in this book. Focusing on delighting users in a meaningful way gives us unique advantage.

The bet Resolute Ventures made on the Sunrise experience was prescient. In 2015, Microsoft acquired the startup for $100 million, and brought many of the insights from the product into Outlook (http://bkaprt.com/dfe2/07-10/).

The lesson we can take away from Sunrise is that designing for emotion from the beginning has a cascading effect. Had Le Van and Valade launched a bare-bones calendaring app without incorporating thoughtful emotional design in their MVP, it's unlikely their startup would have gotten the funding it needed to survive. Additionally, Microsoft probably wouldn't have been interested in acquiring the product if it weren't engaging users in a new and unique way.

With Sunrise, designing for emotion proved to be critical not just to business success, but also to survival.

CONCLUSION

WE'VE COME A LONG way in this little volume, through design and psychology principles applied by Airbnb, Intuit, Headspace, Mailchimp, and Wealthsimple, to name a few. Despite the vast differences in audience, markets, and design, there is a common thread to them all. In each example we see careful consideration for the complex range of emotions humans feel in different phases of the customer journey.

Slack and Airbnb offer us inspiring examples of how we can design more inclusively, and in doing so, create deep emotional engagement while widening our audience. One simple question helps us think inclusively in all of our design processes: Who are we leaving out?

As you begin to design for emotion, keep in mind the timing of your efforts. Design moments that are engaging when your audience is most receptive and when you'll have maximum impact on their experience.

Though functional, reliable, and usable, the examples we've examined go further to create a pleasurable experience. Emotional design connects with an audience in ways we could have never fathomed during the bygone era of usable but unremarkable websites and products. We can channel personality into our work so our users can feel like they're interacting with a human—not a corporate avatar. They appreciate us for our sincerity, and they trust us because they see themselves in our brand. And when we make inevitable mistakes, they'll be more likely to forgive us because our earnestness is visible.

A common thread weaves through the principles and examples in this book: as we reflect on the complicated emotions people bring to our work and as we thoughtfully design for them, we give those we serve a sense of our shared humanity that can bring us closer to the hopeful vision we once had for the web.

We're not just designing screens. We're designing human experiences. Like the visionaries of the Arts and Crafts movement, we know that preserving the human touch and showing ourselves in our work isn't optional: *it's essential.*

ACKNOWLEDGMENTS

THIS BOOK WOULD NOT be resting in your hands if not for the generous opportunity afforded me by Jeffrey Zeldman, Mandy Brown, and Jason Santa Maria. Katel LeDû planted the idea of a second edition of this book in my mind and has been a great supporter along the way. I'm honored to be the purple stripe in A Book Apart's "rainbow of knowledge" and humbled to be part of such a brilliant lineup of authors.

I owe a great deal of thanks to Mandy Brown and Krista Stevens, who made me look like a better writer than I am in the first edition of this book, and Sally Kerrigan will forever have my gratitude for helping me find my way in this second edition. They were patient with me when I bumbled through tricky passages and kindly offered encouragement right when I needed it. Mandy, Krista, and Sally, you made writing this book fun, and for that, I thank you.

Whitney Hess shared invaluable feedback on early drafts of the first edition that kept me pointed in the right direction. I'm grateful to have had her expert opinion and eagle eye on this book.

I'm also grateful that my pal Jared Spool wrote such a lovely forward to this book, but more importantly, he saved me from blowing off the side of a hill in Vik, Iceland. That was a close one!

I spent a lot of time researching for this book, but I had some indispensable help from the Twitterverse. Thanks all who answered my late-night calls for help on Twitter. Tina Roth Eisenberg, you delivered some real gems!

Thank you, Rudy Adler, Kara DeFrais, Adam Glynn-Finnegan, and Sarah Parmenter, for taking the time to talk through the ideas in this book. Your stories and insights were invaluable.

Writing a book is hard work, and it helps to have a support system to keep your spirits high. My wife Jamie shouldered parenting duties to give me time to write and encouraged me when I needed it the most. It's extraordinary to have someone so generous in my life.

Last but not least, I am the father to two amazing boys, Olivier and Bellamy, who have taught me more about emotion than I ever thought possible. I'm lucky to be their dad.

RESOURCES

DESIGN PRINCIPLES ARE AN essential guide as we solve visual and conceptual problems. Let's face it, if you don't have a solid foundation in basic design principles, you're destined to succumb to the temptations of decoration and design without meaning. If you can buy but one design book, make it *Universal Principles of Design* by William Lidwell, Kritina Holden, and Jill Butler.

I'm certainly not the first person to see the connection between design and emotion. The legendary design thinker, Donald Norman, searches for the reasons we fall in love with products in his timeless book *Emotional Design.*

There's a lot of science and psychology behind the concepts and techniques we explored in this book. If you want to dig deeper still into the fascinating magic happening between our ears, start with Steven Pinker's *How the Mind Works, The Blank Slate: The Modern Denial of Human Nature,* and Antonio Damasio's *Descartes' Error.*

Youngme Moon's book *Different* takes a look at why some brands manage to stand out and become market leaders.

Dr. John Medina provides a great deal of insight into the relationships between the way our brains are structured and the way we behave. You'll find his book *Brain Rules* an interesting read.

Sara Blakely, founder of Spanx, has a natural gift for designing for emotion. I highly recommend her Master Class video series (http://bkaprt.com/dfe2/08-01/).

I've referenced user research a number of times in this book. To sharpen those skills, I recommend you turn to Erika Hall's book Just Enough Research.

Alan Klement's article A Script to Kickstart Interviews About Someone's Jobs to be Done (http://bkaprt.com/dfe2/08-02/) is a worthwhile read that will give you a specific script to follow when talking to customers about why they bought (or abandoned) your product.

As we design for emotion, we need to consider how a code of ethics can keep us pointed in the right direction. Mike Monteiro's design ethics offers a solid foundation on which to build (http://bkaprt.com/dfe2/08-03/).

As we saw in Chapter 7, designing for emotion is easily folded into the design sprint process. If you're new to running design sprints, consult Richard Banfield's ebook Enterprise Design Sprints (http://bkaprt.com/dfe2/08-04/).

You'll find a host of compelling articles exploring psychology, emotion, and user experience around the web. Here are a few of my favorites:

- Jared Spool, "Understanding the Kano Model – A Tool for Sophisticated Designers" (http://bkaprt.com/dfe2/08-05/)
- Interaction Design Foundation, "Minimum Viable Product (MVP) and Design – Balancing Risk to Gain Reward" (http://bkaprt.com/dfe2/08-06/)
- Dana Chisnell, "Beyond Frustration" (http://bkaprt.com/dfe2/08-07/)
- Susan Weinschenk, "The Psychologist's View of UX Design" (http://bkaprt.com/dfe2/08-08/)
- Nathanael Boehm, "Organized Approach to Emotional Response Testing" (http://bkaprt.com/dfe2/08-09/)
- Trevor van Gorp looks at the role of personality in emotional design in his article "Emotional Design with A.C.T.: Part 1" (http://bkaprt.com/dfe2/08-10/).

REFERENCES

Shortened URLs are numbered sequentially; the related long URLs are listed below for reference.

Chapter 1

01-01 https://aneventapart.com/news/post/practical-branding-by-sarah-parmenter-an-event-apart-video

01-02 https://en.wikipedia.org/wiki/Facebook%E2%80%93Cambridge_Analytica_data_scandal

01-03 https://www.wired.com/story/five-years-tech-diversity-reports-little-progress/

01-04 https://hbr.org/2015/01/intuits-ceo-on-building-a-design-driven-company

01-05 https://medium.com/@GarronEngstrom/principles-of-emotional-design-c79d40c05d70

Chapter 2

02-01 http://darwin-online.org.uk/content/frameset?itemID=F1142&viewtype=text&pageseq=1

02-02 https://faculty.uca.edu/benw/biol4415/papers/Mickey.pdf

02-03 https://en.wikipedia.org/wiki/Golden_ratio

02-04 https://en.wikipedia.org/wiki/Vitruvian_Man

02-05 https://muledesign.com/2017/07/a-designers-code-of-ethics

Chapter 3

03-01 https://www.flickr.com/photos/clagnut/4947389773

03-02 https://en.wikipedia.org/wiki/File:Bundesarchiv_Bild_146II-732,_Erholung_am_Flussufer.jpg

03-03 https://www.youtube.com/watch?v=1rV-dbDMS18

03-04 https://www.nngroup.com/articles/persona/

03-05 https://www.nngroup.com/videos/why-personas-fail/

03-06 https://styleguide.mailchimp.com/

03-07 http://aarronwalter.com/design-personas/

03-08 http://significantobjects.com/about/

03-09 https://lectureinprogress.com/journal/anna-charity

03-10 https://api.slack.com/best-practices/voice-and-tone

Chapter 4

04-01 https://thecreativeindependent.com/guides/how-to-begin-design-ing-for-diversity/

04-02 http://www.benedelman.org/publications/airbnb-guest-discrimina-tion-2016-09-16.pdf

04-03 https://open.nytimes.com/diversity-inclusion-and-culture-steps-for-build-ing-great-teams-ca157bd98c07

04-04 https://www.microsoft.com/design/inclusive/

04-05 https://medium.com/@uxdiogenes/just-a-brown-hand-313db35230c5

04-06 https://airbnb.design/your-face-here/

Chapter 5

05-01 https://www.atlassian.com/team-playbook/plays/customer-jour-ney-mapping

05-02 https://www.youtube.com/watch?v=XgRlrBl-7Yg

05-03 https://www.invisionapp.com/inside-design/oral-history-of-mailchimp-high-five/

05-04 http://stanford.edu/~gbower/1969/Narrative_stories.pdf

05-05 https://greatergood.berkeley.edu/article/item/how_stories_change_brain

05-06 https://www.patagonia.com/blog/

05-07 https://www.patagonia.com/artifishal.html

Chapter 6

06-01 https://www.wealthsimple.com/fr-ca/magazine/news-redesign

06-02 http://airbnb.com/trust

06-03 https://www.flickr.com/photos/14922438@N00/194463892/

06-04 https://www.flickr.com/photos/41225983@N00/193706751/

06-05 https://blog.flickr.net/en/2006/07/19/temporary-storage-glitch/

06-06 http://interactions.acm.org/content/?p=1226

Chapter 7

07-01 https://labs.sogeti.com/the-minimum-lovable-product/

07-02 https://en.wikipedia.org/wiki/Agile_software_development

07-03 https://twitter.com/jopas/status/515301088660959233?lang=en

07-04 https://www.inc.com/magazine/201802/mailchimp-company-of-the-year-2017.html

07-05 https://medium.com/@awilkinson/slack-s-2-8-billion-dollar-secret-sauce-5c5ec7117908

07-06 https://www.youtube.com/watch?v=QdOPL-2SkmI

07-07 https://www.atlassian.com/blog/announcements/introducing-stride

07-08 https://www.atlassian.com/blog/announcements/new-atlas-sian-slack-partnership

07-09 http://resolute.vc/blog/2013/06/backing-sunrise/

07-10 https://techcrunch.com/2015/02/04/microsoft-sunrise/

Resources

08-01 http://universalprinciplesofdesign.com/books/

08-02 https://www.goodreads.com/book/show/841.Emotional_Design

08-03 https://stevenpinker.com/publications/how-mind-works

08-04 https://www.penguinrandomhouse.com/books/290730/the-blank-slate-by-steven-pinker/

08-05 https://www.penguinrandomhouse.com/books/297609/descartes-er-ror-by-antonio-damasio/

08-06 https://www.penguinrandomhouse.com/books/200292/different-by-youngme-moon/

08-07 https://www.goodreads.com/book/show/2251306.Brain_Rules

08-08 https://www.masterclass.com/classes/sara-blakely-teaches-self-made-en-trepreneurship

08-09 https://abookapart.com/products/just-enough-research

08-10 https://jtbd.info/a-script-to-kickstart-your-jobs-to-be-done-interviews-2768164761d7

08-11 https://github.com/mmmonteiro/designethics

08-12 https://www.designbetter.co/enterprise-design-sprints

08-13 https://articles.uie.com/kano_model/

08-14 https://www.interaction-design.org/literature/article/minimum-viable-product-mvp-and-design-balancing-risk-to-gain-reward

08-15 https://uxmag.com/articles/beyond-frustration-three-levels-of-happy-design

08-16 https://uxmag.com/articles/the-psychologists-view-of-ux-design

08-17 https://uxmag.com/articles/organized-approach-to-emotional-re-sponse-testing

08-18 https://boxesandarrows.com/emotional-design-with-a-c-t-part-1/

REFERENCES **105**

INDEX

ABOUT A BOOK APART

We cover the emerging and essential topics in web design and development with style, clarity, and above all, brevity—because working designer-developers can't afford to waste time.

COLOPHON

The text is set in FF Yoga and its companion, FF Yoga Sans, both by Xavier Dupré. Headlines and cover are set in Titling Gothic by David Berlow.

This book was printed in the United States using FSC certified papers.

FSC
www.fsc.org

ABOUT THE AUTHOR

Aarron Walter is VP of design publishing at InVision, drawing upon fifteen years of experience running product teams and teaching design to help companies enact design best practices. Aarron founded the UX practice at MailChimp and helped grow the product from a few thousand users to more than 10 million. His design guidance has helped the White House, the US Department of State, and dozens of major corporations, startups and venture capitalist firms. You'll find Aarron sharing thoughts about design on Twitter @aarron, and as the cohost of the Webby nominated Design Better podcast.